P9-AZX-455

DISCARDED
FROM
UNH LIBRARY

Quality Assurance and TQM for Analytical Laboratories

Quality Assurance and TQM for Analytical Laboratories

Edited by

M. Parkany
International Organization for Standardization
Geneva

The Proceedings of the Sixth International Symposium on the Harmonization of the Role of Laboratory Quality Assurance in Relation to Total Quality Management (TQM), held December 1995, Melbourne, Australia.

Special Publication No. 169

ISBN 0-85404-760-3

A catalogue record for this book is available from the British Library

Published by The Royal Society of Chemistry,
Thomas Graham House, The Science Park, Milton Road,
Cambridge CB4 4WF, UK

Printed and bound by Athenaeum Press Ltd, Gateshead, Tyne and Wear, UK

Foreword

by Dr. Lawrence D. Eicher, Secretary-General, ISO
(International Organization for Standardization)

The success of the Proceedings of the Fifth ISO/IUPAC/AOAC Symposium in 1993 on "Quality assurance for analytical laboratories" can be attributed to the volume of excellent papers which present the state of the art of the relevant technology at the time.

I noted one paper, however, which was different, and this makes the bridge between that Symposium and the present one. I refer to "Management and motivation in quality assurance: The human element in proficiency", a paper given by Ms. T.K. Blumenthal, which addresses a crucial component of total quality management (TQM). This present volume includes a definition of TQM as well as lectures on the subject. I am therefore not trying here to give "my" definition of TQM for laboratories; I would rather give some illustrations - a sort of caricature.

We may certainly agree that laboratory analysis is not an isolated action but is, in most cases, a service for others, using samples taken, reagents and instruments prepared or supplied. Within the laboratory, a team of individuals performs the tasks. They are expected to use distilled water (not tap water), specified reagents, calibrated instruments, the correct procedure, etc. These are all indispensable elements of the quality control process, which is part of quality assurance. Total quality management contains these elements, and more.

It also contains the human element - the devotion, the culture .- and even more. In fact, the boundaries of TQM cannot be fixed or closed. More and more new elements will be discovered that will enrich TQM, one important aspect being the continuous improvement of human activity in all directions.

In clinical laboratories, there are evident TQM "musts" - for example, not to confuse the samples, not to use contaminated needles for taking blood samples, etc. The analysis should be carried out correctly and as quickly as possible. Results should be conveyed immediately.

Further, when people in a clinical laboratory strive to detect problems much earlier, to find faster and more accurate methods, to perform the analysis from a reduced blood sample - they are already approaching TQM.

It is very easy to estimate the status of TQM in any laboratory: ask the personnel (from the head of the laboratory to the assistants) about motivation in their work and whether they are satisfied with their results.

We hope that these lectures will give you new ideas and will open new horizons for further improvement.

Introduction

by Dr. Michael Parkany, Senior Technical Officer, ISO (International Organization for Standardization)

In the series of the International Symposia for Harmonization in the field of Analytical Chemistry supported by the International Council of Scientific Unions (ICSU) this is the sixth one.

At this time ISO, IUPAC (International Union of Pure and Applied Chemistry) and NATA (National Association of Testing Authorities, Australia) have joined their efforts to bring together in Australia distinguished specialists of Quality Assurance and Laboratory Management, to listen to about thirty lectures on this subject and to have discussion on them.

NATA (having offices in Melbourne and in Sydney) has kindly accepted the task of organizing the Symposium there. I wish to acknowledge the hard work of all local organizers and particularly:

Dr. Helen Liddy (Melbourne) and
Mr. Paul Davies (Sydney)

They have contacted Australian specialists, discussed with them the subject and selected the lecturers. They have helped and encouraged them to contribute to the Symposium.

ISO for its part has contributed to the Symposium with logistic help (e.g. printing and mailing the information leaflets, giving publicity to the Symposium) and also supplying the selected list of the relevant ISO International Standards and ISO/IEC Guides.

I am pleased to note that this is the first time that Australian colleagues participate in such an impressive number at our International Symposium on QA and TQM for Analytical Laboratories and present their achievements at this international forum.

Studying in depth the lectures, the first striking impression is the unanimous positive experiences of laboratories that had relied on the services of consultant firms performing their external auditing as well as those that have made extensive use of Certified Reference Materials (CRMs) and those participating in interlaboratory proficiency testing.

Consultant firms have not only the indispensable knowledge but also the experience how to help to overcome difficulties in a relatively much shorter time compared with laboratories that want to rely entirely on their own experience.

A further important aspect is that producers of analytical instruments put on the market intelligent "measuring systems" instead of the classical measuring instruments. The "measuring systems" do a lot for the analytical chemist in calibration, internal control, statistical analysis, decision, documentation, etc. They are clever, organic parts of the lab's Q.A. system.

Finally, I wish to acknowledge the help of the active members of the IUPAC Interdivisional Working Party on Harmonization of Quality Assurance Schemes for Analytical Laboratories.

We all extend our thanks to the Authors providing their manuscripts and to Mrs. J.A. Seakins of the Royal Society of Chemistry for arranging this special publication.

Contents

Motivation and 'Marketing' of the Analytical Laboratory for TQM: Pride of Place

Trean Korbelak Blumenthal

LIBRA TECHNOLOGIES, INC., 16 PEARL STREET, METUCHEN, NJ 08840, USA

Total Quality Management (TQM) in an organization includes several key concepts: customer satisfaction; teamwork; individual accountability; and continuing quality improvement. ISO 8402:1994, 3.7 defines the concept of quality in TQM in terms of "the achievement of all managerial objectives." Among analytical laboratory managerial objectives are the desire for recognition and status / prestige for the laboratory as a true partner in the larger organization it serves and the ongoing motivation of the lab staff to continuing development and confident performance at productive and high quality levels. How can we achieve customer satisfaction and motivate analytical laboratory personnel to be part of a total, and continuing, quality management effort, both within the lab and between the lab and the larger organization in which the lab resides?

The analytical laboratory all too often is an undervalued asset which is not granted the visibility and recognition it deserves for the vital, and often central, role it plays in organizations. I have seen, in both industrial and government settings, the analytical laboratory treated with everything from benign neglect, through grudging tolerance, all the way to outright hostility and resentment. This is astonishing when one considers the countless important decisions made which depend entirely upon the results of measurement processes, and for which the quality of the decision, and its attendant consequences, are directly linked to the quality of analytical lab data and the integrity, skill, and knowledge of lab scientists and managers.

This behavior on the part of the lab's "customers", i.e., users of the lab's "products" --- results, interpretations and reports --- is counterproductive to the well-being of the organization. People do not do their best work when they perceive that their work is not valued. A lack of understanding and appreciation of the contributions of the analytical lab is very demotivating to its members. In turn, this can contribute to less than full effectiveness in lab performance, which then results in less than optimal product going to the customer, and so the cycle goes, with diminishing confidence on both sides.

In an effective TQM organization, there must be mutual understanding and respect among the partners in the enterprise, in order to achieve the required communications and teamwork. Rather than being taken for granted, or considered little more than a necessary (and costly) evil, the analytical lab needs to be brought to its rightful place as a valued and essential partner, with visibility and recognition for the value of its products. In short, the analytical lab and its products, and the analytical / measurement sciences themselves, need **marketing**, just as a business markets the enterprise (to its investors) and the products of the enterprise (to its customers). The lab must realize and make visible **what** it does; what it **takes** to do it; what it **means**; and **why it matters**, and must communicate these things effectively, to both its customers and the upper management of the organization. In turn, the lab must listen carefully to its customers and its upper management, to understand their needs, their issues, and their language and vocabulary. The lab needs further to consider how it can best meet those needs, and communicate that it has done so, to achieve customer satisfaction, while at the same time exercising its unique skills and knowledge to contribute to the well-being of the entire organization and <u>its</u> customers. Finally, the lab needs a feedback mechanism to assure that the customer is satisfied, and /or to find ways to do better.

As analytical scientists, we may feel that the vital work we do is self-evident in its worth: but this is not necessarily so, and it profits us to learn how to do more effective promotion of the lab and improve its image. Lab personnel will be more highly motivated to excellence as the prestige of the lab and its products rises and their individual sense of pride of place increases.

One approach to "marketing" the analytical lab is to consider it a business unit within the larger enterprise. It doesn't matter whether the larger enterprise is itself an industrial business concern or not: the same principles can apply to a government organization, an academic institution, or a research or other type of institute. The "department as a business" orientation can give profound new insights into the successful operations of a lab, and is extremely useful in terms of communicating economic impact to upper management in language they understand and appreciate.

For a business to function effectively, it must have, among other things:

- a clear idea of its **mission**, shared by all members of the group
- a form of **organization** which supports that mission
- defined **products** / services appropriate to its mission
- an understanding of its **customers** / market
- **communications** pathways to / from its customers / market
- **standards** for performance
- economic knowledge and control: an "**accounting**" system
- an **inventory** of stock-in-trade: specialized methods, versatile equipment, knowledge bank, vital records, databases

<u>Mission</u>

It is very important that the laboratory manager carefully define the mission of the lab, i.e., the reason(s) for the existence of the lab unit in the organization. Understanding the mission gives power and meaning to the managerial organization of the lab, both in terms of physical plant and personnel, and to setting and managing the goals, priorities, and strategies for success of the lab business unit. Furthermore, it is equally important that every person who works in the lab share a clear view of the lab's mission and "buy into" it. This gives a context to the work which is very empowering and keeps everyone working in the same direction and perceiving themselves as part of a team with a common purpose. The TQM concept then extends this team view to the larger organization and enables more effective interaction among groups, each of which sees its unique and important place in the total enterprise.

In a TQM environment, a long view is helpful. For example, if an analytical lab defines its mission as no more than that of making measurements and reporting results, there may result a perception, and corresponding devaluation, of these activities as no more than a technician's work, no matter how complex and demanding the effort to produce those measurement results and reports. It is better for an analytical lab to define and communicate its mission as that of **providing scientific evidence for important decisions and the resulting benefits**, such as:

- the fate of materials and products (pass/fail; rework; recycle, etc.)
- properties of materials toward their optimal usage (e.g., reference materials)
- the composition of a previously unknown substance (new knowledge)
- the outcome of a research investigation (what have we / have we not learned)
- the health or illness of a patient
- whether or not a regulation has been violated (support for legal actions)
- whether or not a crime was committed, and how (forensic evidence)
- resolution of a customer problem (outside customer or the lab's "customer")
- the probability of success or failure of a scientific study
- establishing the economic viability of a product in the marketplace (stability)
- whether or not the enterprise should invest in a process, product, or project
- the suitability of a vendor / supplier
- the best way to make a particular measurement
- the validity of a process / method
- the appropriate specification boundaries for a parameter
- gaining better control and consistency of a manufacturing process
- support for or against a new regulation
- how to achieve lowered legal or ethical or community risk

No doubt the reader will be able to imagine many more, and the selection will, of course, depend upon the enterprise the lab serves. These mission viewpoints remove the analytical lab from its isolation as a generator of numbers, and place the products of the lab squarely in the center of much of the corporate (enterprise's) "action". The longer view envisions the lab as part of, and partner to, many other groups in the enterprise and immediately repositions the image of the lab in the eyes of its customers. A consultancy / advocacy approach is a more positive and proactive role for the lab, which in turn enhances the prestige of the lab and of those who work there.

My first professional job was that of analytical chemist for the U.S. Food and Drug Administration. There was no question that we understood the mission of that organization and its duty to protect the public health and safety with regard to the products under its jurisdiction. That knowledge informed all of our actions. We knew that each analysis had significant consequences, however the result turned out. We knew that at any time we might have to go to court and give testimony about our findings (and sometimes we did so), and that our professional reputation and credibility were on the line. We understood that we carried serious responsibilities, both to the public and to the companies whose products we analyzed, to produce a correct and timely analytical result, and we were fully accountable for the quality of our work. We felt ourselves part of an important enterprise. And the interesting thing is, we fully **enjoyed** our work, even when some of the specific jobs were difficult, dangerous, or distasteful. We knew we were making a contribution, and that made all the difference.

Customers

Who are the "customers" of the analytical lab? What are the lab business unit's "sources of revenue?" Who pays for the analytical lab? It depends upon the specific mission of the particular lab and its role within a given organization. For instance, a quality control lab in a pharmaceutical company might have, among its customers, several manufacturing production groups; certain R&D groups; perhaps an International QC group; and the QC/QA groups who use the data for decisions and insights, for example, batch release; statistical analysis for SPC (statistical process control); and stability studies and expiration dating (all of which have very profound economic impact on a company). In addition, the lab may certify reference standards for use by customers both within and outside the organization. Vendor certification programs make the Purchasing department a customer for the lab. What's more, the lab is its own customer, for the materials it certifies, reagents it standardizes, instruments it calibrates, and methods it validates all benefit its own business success by making it a productive producer of quality products.

What do customers want from an analytical lab?

- timeliness
- reliability
- clarity of communication of the results and their meaning
- proper documentation
- cost-effectiveness
- the "good news"

This last can be problematic. Customers may define "good news" as the outcome they had hoped for. Labs cannot always report hoped-for outcomes. Diplomacy and context in the delivery of the news are essential, as are an understanding of the consequences to the customer of various outcomes, and a sensitivity to the customer's ego. One might ask how to go about reporting an objective and well-established scientific result, which, however, is not the hoped-for outcome, without putting oneself in the position of being the messenger who is killed because the message is unpopular. Building a successful TQM environment is one pathway, because if the lab is well-respected and it is understood that even when things go wrong (as revealed by a correct lab analysis), something important has been learned that will have longer-term benefits, the perspective changes to a shared problem-solving approach. For example, if a production manager's compensation is tied to the amount of acceptable product delivered into inventory, that manager will naturally be upset if a batch fails to pass specification, and even more upset if SPC control charting reveals that excess process variability exists. It is not diplomatic to report that the process is "out of control". It is more constructive, while being equally factual, to point out the very large proportion of successes, while helping to diagnose how the batch failed or the excess variability occurred and how to prevent it in the future.

Economics

In these cost-conscious times, lab managers and staff are under many economic pressures, and it is more important than ever that the products of the analytical lab be presented not only in terms of their scientific reliability, but also for their economic value. Lab managers are often familiar with creating budgets, tracking expenses, and justifying purchases, but may be less familiar with a more comprehensive management analysis which takes into account the value to the enterprise of the products of the lab. Analytical lab data support such benefits to the larger organization as reduced waste; liability / risk avoidance or reduction; regulatory compliance assurance; contribution to profit (e.g., support for a new patent or a new product); cost savings; increased efficiency and productivity; improved customer service; competitive advantage (e.g., technical data the Marketing department can use to show product superiority); newsworthy discoveries benefiting a larger community; and so forth, but are rarely presented, or even considered, in that light.

In order to gain insight into the quantitative economic value of the lab's work, it may be necessary for the lab manager to get out among his or her peers in the "customer" departments and begin learning more about the overall enterprise in economic terms, and how the results from the lab relate to it. It is also important to track the time and cost of lab analyses and projects, and to analyze periodically what proportion of the lab's work supports each of its customers. Even a good estimate can be very revealing. If one then compares the benefit to the organization versus the cost, one can gain knowledge useful for setting priorities and managing for maximum value. In some organizations, an administrative group performs cost analyses in order to allocate the cost of lab operations in a chargeback to the departments the lab supports. If such information exists in your organization, it can be very useful to you, so it pays to find out if it is available.

In a TQM environment, the lab manager gets involved in the economic impact of his/her operation, and thereby becomes better positioned to negotiate for personnel, equipment, and other needs that may previously have been viewed only as expenses, but now can be seen as investments. Economics is the language of upper management, and the lab that reports in those terms wins increasing respect and support.

Stories from the Battlefield

Business has been described as war, and here are two stories, from the battlefield, of successful application of the principles described in this paper:

When I worked in management in a large pharmaceutical company, my quality assurance group handled a large and very complex workload, and I saw an opportunity for greater efficiency and effectiveness in the accomplishment of our mission through the use of microcomputer technology. This was more than ten years ago, when microcomputers were new on the scene. In spite of political barriers to this purchase, it was possible to justify it based on an analysis of cost savings to the larger corporate enterprise which were expected to result from the data handling and more comprehensive information reporting capabilities we would gain. These were expressed in terms of the quantitative value of that information to our customers' decision-making power. (Our customers were Manufacturing; R&D: product, process, specification, packaging development; Purchasing; and Regulatory Affairs, among others). The return-on-investment analysis was done conservatively and it was predicted that the computer system would pay for itself in one year. In fact, it paid for itself in only six months, and there were other benefits in terms of development of the staff's workload management capabilities, supervisory skills, and general confidence and empowerment over their areas of responsibility, not to mention the enthusiastic energy with which they learned and advanced the applications of the technology for even greater benefits than predicted.

The second story concerns a product stability (shelf-life) program and interdepartmental communications. Pharmaceutical products are monitored to assure that manufactured lots in their commercial packaging exhibit stability at least comparable to the product as initially approved for sale by the government. A stability program is a large and complex activity, involving drawing samples from production lots; storing them in special facilities under controlled conditions for defined time intervals; "pulling" them for analysis at the right times; conducting lab analyses; compiling and analyzing the data; and producing timely reports. My group was responsible for everything but the lab analyses themselves. When there are many products, batches, and package sizes, the workload impact is large, and the proportion of the analytical lab's time spent on this program is large. The program therefore can be quite costly, but, aside from the fact that it is required by law, there are economic benefits to the corporation in that the knowledge gained serves to allow the extension of expiration dates, better new product and package development, and so forth. Widespread misunderstanding on the part of other departments about this program led us to the idea of presenting a series of seminars to describe and explain it. In the spirit of TQM, each person in my group who played a role in the program was invited to speak about their work. One or two of my staff were shy about public speaking. They were told they would not be required to participate, but that no one could be better qualified to describe the work they did. In a surprising move, the most reluctant speaker, who was an artist in her free time, made a wonderful and creative display which illustrated her work beautifully, and came forward to give an excellent talk. Others also rose to the occasion and gave outstanding presentations, and the seminars were so successful that departments not initially included were requesting the seminar for their groups. Not only did this effort improve understanding and communications about this important program, but the prestige and confidence of my group as individuals and as a team skyrocketed, and the visibility and appreciation for their work was gratifying and highly motivating.

In Conclusion

Total Quality Management offers an opportunity for the analytical laboratory to be re-positioned as a high-status partner in an enterprise which fosters teamwork and cooperation. A position of higher prestige enables the lab to attract and retain superior staff and to motivate them to higher achievement through pride in identification with the group and a feeling of *esprit de corps*. The analytical lab can also win greater support for advanced tools and technologies which improve further its products. The highest technologies are often used first in analytical settings, e.g., robotics, LIMS (lab information management systems [computerized]), radiotracers, mass spectrometry, computer-assisted workstations, etc. The laboratory can often be an early adopter of high technology, such as microcomputers, thus paving the way for introduction throughout the organization with a cascade of benefits or "profit" to that organization.

There is a need and benefit as well in fostering greater prestige for the analytical and measurement sciences in universities and colleges, so that good students will be attracted and prepared for advanced study and quality employment. Perhaps analytical scientists need to do more to promote and market the profession.

Finally, there is a need for the accounting disciplines to work together with scientists to address, in a more meaningful way, the asset value of science, and, in particular, analytical lab products, so that the ledgers reflect not just an expense, but a balance and a positive value.

Quality in the Analytical Laboratory

Bernard King

LABORATORY OF THE GOVERNMENT CHEMIST, QUEENS ROAD, TEDDINGTON, MIDDLESEX TW11 0LY, UK

1 INTRODUCTION

Whilst QA is rightly high on all our agendas we would do well to be sceptical about some of what is sold as quality management. We need to adopt a total quality approach which places emphasis on good science, good management and empowerment of the analyst. Attention must be focused on the primary issues which determine quality and not on frills or secondary factors. For example documentation is essential but the paper mountain is an expensive diversion. QA only works when it is owned and controlled by the analyst and when it facilitates quality rather than tries to impose simplistic rules.

This paper reports on some other recent work and thinking emerging from activities such as the UK VAM initiative (4), CITAC (1) and EURACHEM (2). Underlying the paper is the author's view that there is more than one way of ensuring quality but that we have not yet got it right (3-5). Some important quality issues which are often overlooked, such as defining the requirement, qualitative analysis, non-routine analysis and R&D are also discussed. The absence of discussion of issues such as traceability (2,5), proficiency testing (9), measurement uncertainty (8), and education and training (12) does not imply that they are not considered important. On the contrary, these aspects of QA are vital. They are not covered here because they are well covered in other places.

2 WHAT IS QUALITY?

The concept of quality is often described by phrases such as "satisfying customers' needs", "fitness for purpose" or "getting it right first time". Whilst these phrases encapsulate the essence of quality, we cannot make progress until we are able to describe in detail what is required. Defined quality criteria enable judgments to be made against defined minimum standards or on a ranking basis. An element of scientific excellence is involved but there must also be a trade off involving cost and time. We have a variety of sources of evidence about the quality of a laboratory including:

- scientific reputation/ability to win contracts
- work practices
- staff qualifications, experience, attitudes
- achievement of performance targets
- quality systems, audit, review and accreditation.

No one approach is foolproof and there is no substitute for an effective dialogue between the laboratory and its customers.

3 QUALITY MANAGEMENT

One way of describing the process of quality management is shown in Figure 1. It involves three inputs, namely: specification of the requirement, systems and people. Quality is a dynamic issue and the output of the process is continuous improvement. The process must ensure that we do the right experiment as well as doing the experiment right. Systems alone cannot deliver quality. Staff must be trained, involved with the task in such a way that they can contribute their skills and ideas and must be provided with the necessary resources. We need to remember that a motivated person can move mountains. Approaches to quality management range from those that emphasise rules, documented procedures and systems to those which aim to establish a customer centred quality culture which harness the energy and initiative of all staff. Over recent years there has been a move towards the latter total quality approach with emphasis on clarity of mission, improved work processes and staff empowerment.

Figures 2 and 3 compare the main quality management approaches with regard to issues relevant to analytical work. Although simplistic and subjective they illustrate the strengths and weaknesses of the various approaches. ISO9000, ISO25 and GLP are well known but some of the other approaches are briefly described below.

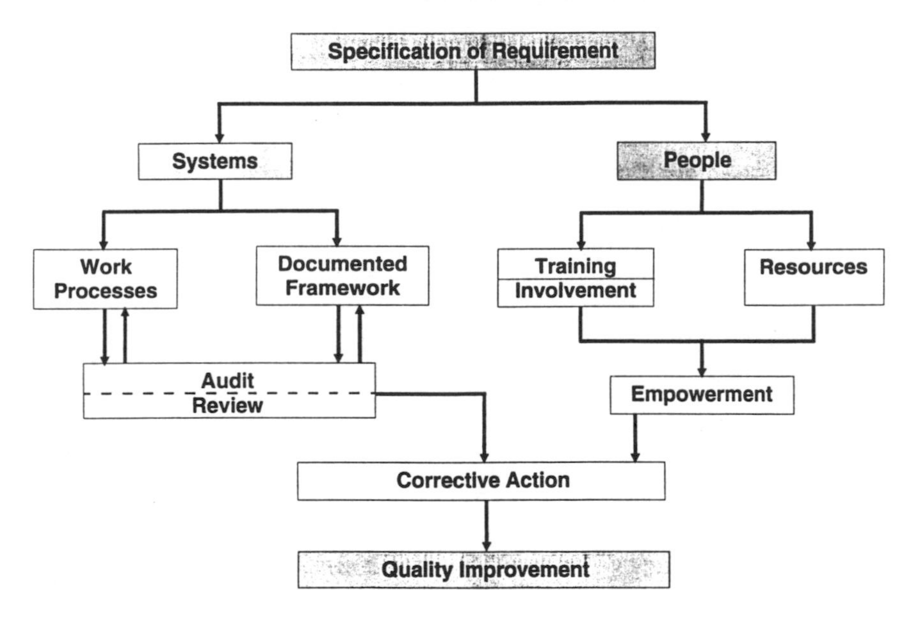

Figure 1 **PROCESS FOR QUALITY MANAGEMENT**

Visiting Groups comprise senior level experts brought together to evaluate a research institute against its stated objectives and with a strong emphasis on the assessment of the excellence of the science, scientists and facilities. Such assessments are most common in academic and other public sector research establishments and are carried out on behalf of customers or funding bodies.

	Standard Available	Creative Thought	Research Design	Experimentation	Interpretation & Reporting	Technical Direction & Project Management
ISO 9000	✓	-	++	++	+	++
ISO 25	✓✓	-	-	+++	+	++
GLP	✓✓	-	+	+++	+	++
Visiting Groups	=	++	++	++	++	++
Benchmarking	=	++	++	++	++	++
TQM	=	++	++	++	++	++
Project Management	=	-	++	++	+	++

KEY

=	No national or international standard	-	Not covered
✓	Flexible standard	+	Covered to some extent
✓✓	Standard with strict criteria	++	Covered
		+++	Strong emphasis

Figure 2 COMPARISON OF DIFFERENT QA APPROACHES

	ISO 9000	**ISO 25**	**GLP**
Specification of task	+	+	+
Information review	-	-	-
Creative thought	-	-	-
Experimental Design	+	-	-
Staff	++	++	++
Equipment	++	++	++
Sampling	+	++	+
Preliminary Analysis	-	-	-
Identification of composition	-	-	-
Quantitative Measurement	++	+++	++
Interpretation/Problem Solving	-	-	-
Reports/Advice	+	+/-	+
Project Management	++	++	++
Technical Direction	+	+	++

Key

- \- = not covered
- \+ = covered to some extent
- ++ = covered
- +++ = covered with emphasis

Figure 3 COMPARISON OF QUALITY STANDARD REQUIREMENTS

Benchmarking is a process whereby an organisation compares its conduct and performance of specific activities with comparable activities in another organisation which is seen to be a market leader but which is not usually a direct competitor. It provides a basis for technology transfer and quality improvement.

Total Quality Management has been extensively applied in Japan and has many variants which are rapidly finding favour in many parts of the world. In addition to the emphasis outlined above TQM is about establishing a leadership style of management which supports and cultivates the one team approach. It accepts risk taking, and recognises that mistakes and failures are inevitable but need to be learnt from. Statistical methods and other tools ranging from brainstorming groups to flow diagrams are used to monitor progress, analyse and solve problems and measure improvement. So-called learning organisations recognise that they are not entirely sure about their strategy and needs, let alone the solutions to all the problems that need to be addressed. Emphasis is therefore placed on continuous learning rather than on rules. TQM is particularly relevant to R&D and is one of the few approaches which address the creative part of laboratory work. The idea of a harmonised standard and third party certification are thought by some to be alien to the basic philosophy but in the author's view TQM underpinned by transparent systems has considerable merit.

Project Management techniques provide a structured approach to work as indicated in Figure 4. This approach recognises that failure often results from inadequate specification of the requirement, poor communication between contributing groups and failure to address problems early in the project. With R&D there is always an element of uncertainty about the feasibility of the project and it is important to structure projects in such a way that key problems are identified and solved before proceeding to full development. Also, it must be recognised that the planning and decision making processes are iterative.

A fuller description and evaluation of the quality management approaches can be found in a report of a study of the QA of R&D (6). It is clear from Figures 2 and 3 that each approach has strengths and weaknesses. ISO 9000, ISO 25 and GLP are reasonably satisfactory for the more clearly defined routine work but less appropriate to R&D when the other approaches have more merit. A hybrid approach would be appropriate for many laboratories.

4 ANALYTICAL QA

At the technical level there are many quality requirements ranging from the validation of methods to the calibration of equipment and the establishment of the traceability of measurements (5). Some of the ingredients of good practice in analytical QA are shown in a mixing bowl in Figure 5. When baking a cake there are some ingredients which are essential, others are optional. A similar situation applies to analytical QA. However, there is not yet agreement on which ingredients are essential and more rigorous studies are required to evaluate the cost benefit of the options. Until we are better able to define an optimum QA strategy, we must avoid attaching ourselves too rigidly to any one approach. However, it is clear that the primary determinant of analytical quality is the skill of the analyst. A trained analyst employing good methods and equipment which has been properly calibrated provides the basis for analytical quality. QA systems, accreditation, benchmarking and QC techniques provide evidence of quality or an indication of problems which need to be addressed. TQM provides a strategy for combining the best of all the approaches.

Analytical work encompasses research, development, non-routine analysis and routine analysis. Although the specific tasks can be very different, the generic subtasks listed in Figure 3 provide a list of topics which must be addressed. The overall quality is only as strong as the weakest link. The quality standards ISO

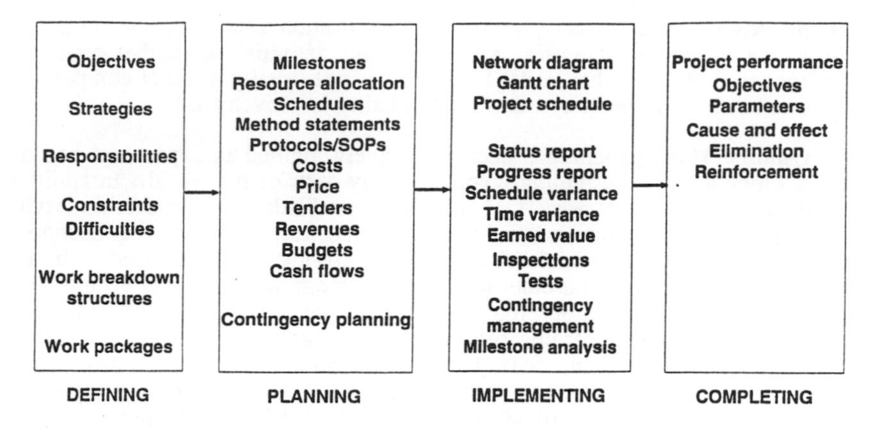

Figure 4 PROJECT MANAGEMENT STAGES

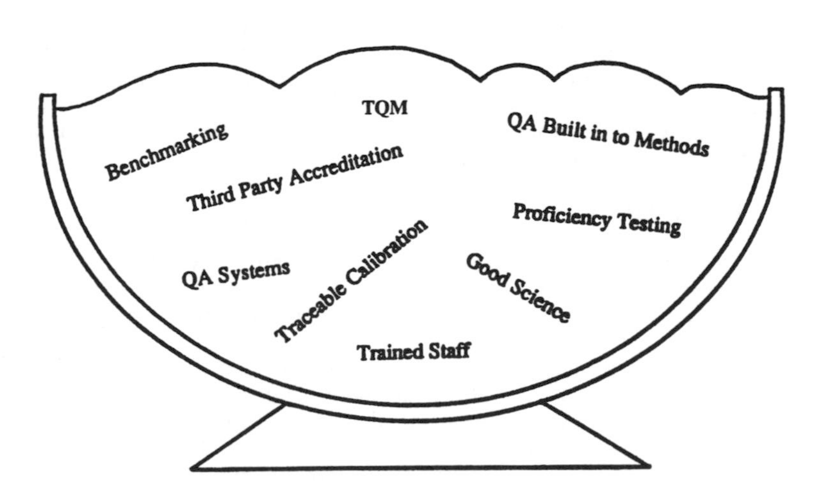

Figure 5 INGREDIENTS OF GOOD ANALYTICAL QA

9000, ISO 25 and GLP together with documents such as the EURACHEM/WELAC guide Accreditation for Chemical Laboratories (7), the EURACHEM guide on Quantifying Uncertainty in Analytical Measurement (8), the draft CITAC International Guide to Quality in Analytical Chemistry - An Aid to Accreditation and others (9,10) provide extensive advice on many aspects of QA but a number of the sub-tasks need to be addressed more fully.

The scale of the analytical task is difficult to gauge and depends on where the boundaries are drawn but is illustrated by a study (11) which estimated the number of laboratories undertaking analytical work in the UK to be 25000 on 11000 sites and with an average number of 9 staff. These estimates include a number of sectors such as engineering laboratories which only undertake a small amount of simple analysis and larger sophisticated sectors such as clinical laboratories which do not normally classify themselves as analytical laboratories. The UK study also assessed the importance analysts attached to quality issues and found the following levels of interest:

	% rating importance as "high"
Use of validated methods	80
Use of reference materials	74
Independent quality audits	60
Proficiency testing	60

Of the 25000 UK laboratories about 500 laboratories have ISO 25 (NAMAS) accreditation or GLP registration for analytical work and many of these are only covered for a part of their total work. To this number must be added an unknown number (< 1000 ?) of laboratories covered by ISO 9000 certification. The inescapable conclusion, however, is that we have a considerable way to go in raising awareness and in attracting laboratories to third party QA systems.

5 ATTITUDES TO QA

Surveys of attitudes carried out as part of the UK VAM initiative show a mixture of contentment and dissatisfaction with third party assessment. UK clinical chemists have set up their own accreditation system based on ISO 25 but with a strong peer group input to both the organisation, assessment and interpretation of quality requirements. The UK ISO 25 market leader NAMAS is well respected with a total of 1,400 laboratories on their books of which 300 undertake analytical work. The reasons for seeking third party registration are partly market driven, particularly in the case of GLP, but are also influenced by the commitment of laboratory managers to quality. The features of third party assessment most appreciated by laboratories are:

- the confidence given to staff, management and customers that quality is under control
- savings resulting from streamlined processes and reduced repeat analysis and customer complaints
- help and advice provided by auditors.

Some of the causes of dissatisfaction are:

- the existence of different schemes (GLP, ISO 25, ISO 9000) which require some laboratories to maintain two or three QA systems. In the UK there is growing pressure for the organisers of schemes to harmonise their requirements or at least establish effective mutual recognition agreements.

- some laboratories have difficulties with the prescriptive interpretation of ISO 25, justified by assessors in the interests of customer confidence. Areas of difficulty include scope statements, documentation of methods, non-routine/R&D work, use of IT and reporting restrictions.

- assessors are sometimes seen as inconsistent, over-concerned with secondary quality issues whilst ignoring other more important issues and on occasions lacking in technical depth. Some laboratory managers appear to be timid of challenging unreasonable demands.

- the requirements of ISO 25 and ISO 9000 are often seen as less appropriate to chemical analysis than physical or engineering testing for which they were first devised. The issues of standard methods, traceability to national standards and calibration are more complex in analytical chemistry. Calibration of the balance or thermometer can be important but for much analytical work these issues are much less important than control of losses during extraction, contamination or interferences. Document control is central to quality in engineering manufacturing but much less important in analytical chemistry.

What are less clear are the attitudes of those laboratories which are not third party registered. From both formal and informal surveys conducted within the UK VAM initiative it is evident that some laboratories are deterred by the cost, others are poorly informed about quality matters whilst others are not persuaded of the added value. A straw poll at a 1993 international meeting concerning the banning of chemical weapons revealed a disappointing attitude to accreditation. Of 12 delegates from around the world only one was from an accredited laboratory, two laboratories were seriously preparing for accreditation, three were preparing slowly and reluctantly, four were positively hostile to accreditation and two had not seriously considered the issue. Despite this overwhelmingly negative attitude to accreditation the laboratories were alert to quality issues and regularly participated in inter-laboratory studies. This cameo picture is consistent with the relatively small proportion of laboratories both in the UK and worldwide which seek third party quality assessment.

Some laboratories see themselves as above accreditation or certification which they regard as only appropriate to routine work. This attitude, of course, misses the point that quality audits are, or should be, peer group review which can add value at any level of sophistication.

The customers of analytical laboratories range from those that understand QA and will only place work with registered laboratories to those who want some form of third party assurance but are not experts and at the other extreme there are customers who have little or no interest in quality issues. Figure 6 shows the results of a survey (9) of customer needs for chemical analysis, whilst Figure 7 shows a worrying level of quality failures. Gaining greater customer interest and input to the analytical quality process must be one of the aims of future work.

6 QA OF REFERENCE MATERIALS

One of the tasks initiated through EURACHEM and CITAC and undertaken by ISO REMCO has been the development of an ISO guide on "Quality System Requirements for Producers of Reference Materials" (draft ISO Guide 34). This document is close to completion but as yet no consensus has emerged about how best to organise a certification system. Given the international nature of RM work it will be important to establish a scheme which is fundamentally international in nature and administered by a body with both QA and technical credibility.

7 SPECIFICATION OF THE ANALYTICAL REQUIREMENT

One of the other neglected issues is specification of the requirement and it is therefore not surprising that uncertainty and concern about quality results. Specifying the requirement is particularly important for investigational work and non-routine analysis. Some of the factors that need to be considered are listed below:

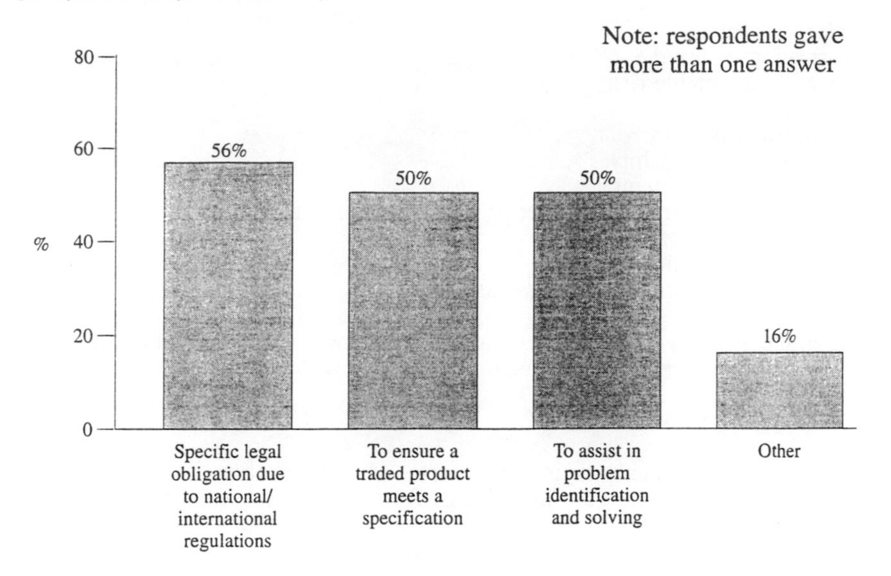

Figure 6 SURVEY OF CUSTOMER REASONS FOR REQUIRING ANALYSIS

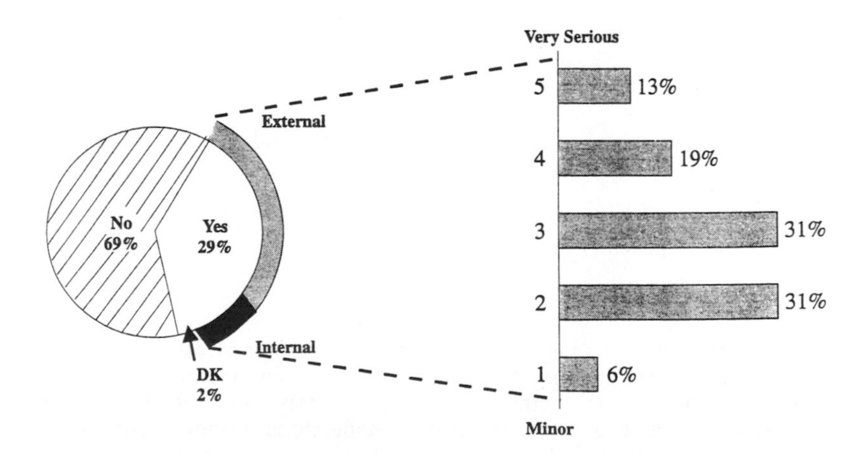

Figure 7 CUSTOMERS WHO HAVE INCURRED LOSS DUE TO POOR ANALYTICAL QA DATA

- analytical context
- information required
- criticality/acceptable risk
- time constraints
- cost constraints
- measurement uncertainty
- traceability requirements
- identification/confirmation/finger printing requirements
- method requirements
- QA/QC requirements
- analytical plan/approval requirements

Even when the answers to these questions are not known it is important to consider the issues and arrive at a pragmatic judgment about how they affect the requirement.

8 QUALITATIVE ANALYSIS

A first step of much analysis is some form of identification, confirmation of composition or finger printing. Tasks range from the analysis of a pure substance to the identification of complex compounds, such as dioxins, at ultra-trace levels in complex matrices. Analytical requirements vary from mass screening with bias towards false positives or negatives depending on the application to highly reliable identification. In all cases the questions are: what level of assurance is required, what level of assurance is provided by a given procedure and how can we develop strategies which provide a national basis for assessing the issues. For example, the courts rightly ask that evidence should be reliable beyond reasonable doubt. What does this mean and how can we measure it? These issues are beginning to be discussed within EURACHEM and CITAC. One way of starting this rather daunting task is to examine the different analytical approaches. For example techniques can be classified as follows:

- Single datum techniques such as melting point, chromatographic retention index or screening test result
- Close match criteria such as MS library searching
- Correlation with experimental data for example, structural elucidation by NMR spectroscopy
- Multi-step structural analysis for example elemental analysis followed by identification of functional groups.

A project has recently been started which will look at both theoretically based logic strategies and empirically based best practice.

9 RESEARCH AND DEVELOPMENT

The QA of routine analysis is reasonably well addressed but the QA of investigational work, ranging from research in to new analytical techniques to the development of a new method has only recently been added to the agenda. Non-routine analysis of reasonably well understood samples constitutes another category of work as does the analysis of totally unknown samples. In all cases investigation, development and validation are required. Some of the factors that need to be considered are:

- Doing the right experiment as well as doing the experiment right
- Nurturing scientific excellence
- Managing the conflict between creativity and rule breaking and orderly planning

and execution of work
- Study plans
- Peer group assessment of the quality and quantity of output
- Fostering communication and cross fertilisation
- Method validation, calibration and QC

A very much more flexible approach to issues such as scope and documentation is required but it would be reasonable, for example, to demand documented strategies for the selection and validation of methods and generic methods for analytical operations.

10 LIMIT SETTING

Analytical data is often used to decide whether a sample complies with regulatory, production or trade requirements which entails comparing results with limit values.

Now that we have strategies for the assessment of analytical uncertainties in sight (8) it raises the question of how we can compare our analytical data which comprise a probability distribution with a single point limit value. Clearly it does not make sense and it will be necessary to also consider the limiting value as a band. Ideally we should be able to devise limits which take account of cost/benefit or integrated risk assessment. The fact is that at present limits are often set on an arbitrary basis which takes more account of politics, bargaining between interested parties, and perceived measurement possibilities than cost/benefit or risk assessment information. In addition many regulations and specifications require the absence of certain toxic or otherwise deleterious substances. Setting a zero limit may have been a pragmatic strategy when limits of detection were measured in parts per million but with modern measurement capability it no longer makes sense. CITAC has agreed to examine how matters may be improved and Figure 8 illustrates some of the issues. In situation 1 the sample is clearly below the limit and the uncertainties are of no consequence to the compliance decision. In example 2 it is not clear cut and we can not escape an assessment which includes probability statements. Reducing the uncertainty does not necessarily alter the decision-making principles.

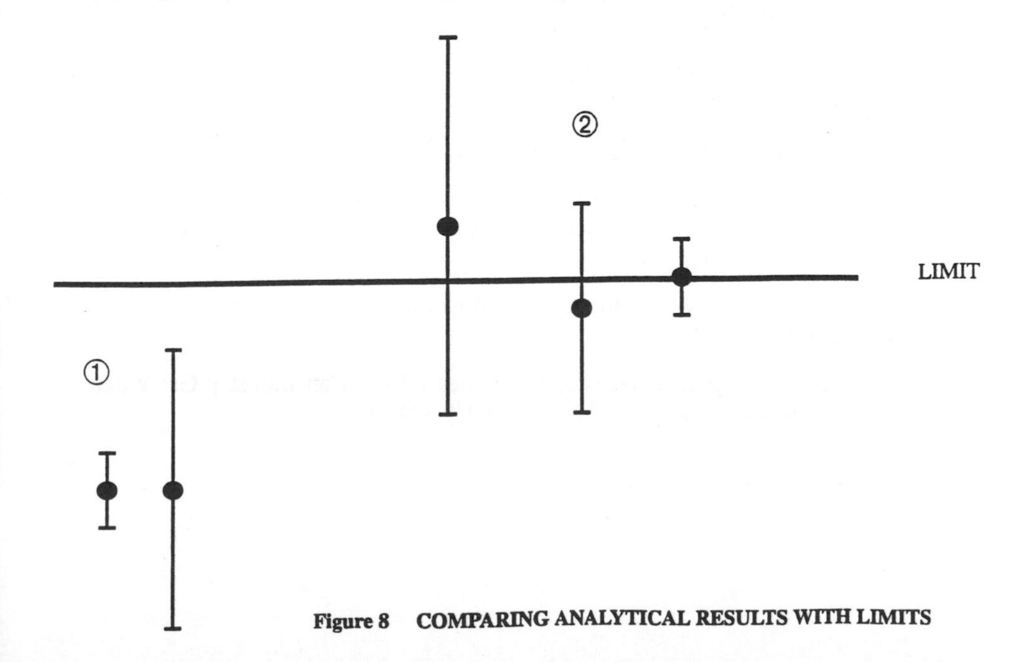

Figure 8 COMPARING ANALYTICAL RESULTS WITH LIMITS

11 CONCLUSIONS

- Third party QA systems are well established and provide a very useful aid to analytical QA but the current approach has failed to attract many laboratories and falls short of the full requirements.
- A hybrid approach which combines the benefits of good QA systems with the outlook and attitudes of TQM has considerable merit.
- The quality of analytical work is only as strong as the weakest link and whole areas of the analytical task are commonly neglected in current QA practice.
- Quality is primarily determined by people and by the application of good science, Quality improvement must therefore plan emphasis on these issues.
- A variety of international organisations are making valuable contributions to quality improvement. Cooperation is vital if we are to make best use of limited resources, ensure effective technology transfers and build a harmonised international chemical measurement system.

REFERENCES

1. B. King and R. Walker, Analytical Chemistry, 1994, 66, 1168A.

2. P. De Bièvre, B. King and W. Wegscheider, Analytical Proceedings, 1994, 31, 377 and 379

3. B. King, Analyst, 1993, 118, 587.

4. B. King, "The VAM Bulletin" (an LGC publication), 1993, 10, 3.

5. Measure for Measure : Recent Developments In The International Chemical Measurement System [In preparation], B. King.

6. B. King and E.A. Prichard, Quality Assurance of Research and Development [In preparation]

7. EURACHEM Guidance Document No 1/WELAC Guidance Document GD2, Edition 1, April 1993.

8. Quantifying Uncertainty in Analytical Measurement, EURACHEM Workshop, Delft, Graz, September 1994.

9. R. Walker, Analytical Proceedings, 1993, 30, 257

10. ISO, IUPAC, AOAC International, Harmonised Guidelines for Internal Quality Control in Analytical Chemistry Laboratories, Pure and Applied Chemistry (to be published in 1995)

11. J. Fleming, Analytical Laboratory Population In the UK, UK VAM Report,November 1993

12. J. Fleming, M. Sargent, Chemistry Graduates - How Can Industry Get What It Wants?, Managing the Modern Laboratory (In Press)

* **Note:** ISO25 refers to ISO Guide 25

Proficiency Testing: A Foundation for Laboratory Quality Improvement

Alan L. Squirrell

NATIONAL ASSOCIATION OF TESTING AUTHORITIES, AUSTRALIA, 7 LEEDS STREET, RHODES, NSW 2138, AUSTRALIA

1. A RISKY BUSINESS

The occurrence of errors in testing and measurement activities is a serious matter for both laboratories and their clients. For the laboratory it can mean expensive retesting, loss of reputation, termination of a contract and poor staff morale. For the client it can mean a wrong decision based on inaccurate test results, failure of components, liability to clients and public, and loss of confidence in the laboratory. One-off errors can be bad enough, but it is the recurring errors that go undetected which can be the most disastrous.

The sources of errors can be diverse and unpredictable. The six key components in laboratory operations: staff, equipment, methods, sampling, data-processing and testing environment, all contribute their share of problems to testing and measurement tasks. At the one time, errors can arise from any or all of these sources. How then can a laboratory hope to detect, deal with and prevent the recurrence of these errors? The answer rests with the implementation of an effective quality assurance system.

2 DEALING WITH ERRORS

A laboratory quality assurance system includes several components which together can search for, pin-point and alert laboratories to errors. The components of a laboratory quality assurance system include:

- Internal Quality Control
- Laboratory Accreditation
- Proficiency Testing

2.1 Internal Quality Control

The nature of the internal quality control program used by a laboratory will vary according to the type of testing or measurement it undertakes. It may involve techniques such as:

- The use of certified reference materials
- The repetitive or replicate testing of samples
- The statistical analysis of results over a period of time

It is important that laboratory managers understand and are able to evaluate their results with at least a basic statistical approach. The ability to analyse quality control data and accumulated results can be a very powerful tool in developing a better understanding and control of the laboratory's testing practices. It also allows the laboratory to develop acceptance ranges for future testing. However, there is a danger in relying on internal quality control alone as a means of detecting and eliminating problems. The laboratory needs to be able to relate its internal quality assurance to the outside world, in particular to other laboratories and to recognised authorities in testing and measurement. This is where the other key components - accreditation and proficiency testing - play a role.

2.2 Accreditation

Laboratory accreditation provides the laboratory with an independent evaluation (by technical experts and quality professionals) and recognition of technical competence to perform specific tests. Accreditation may not be available to (or, in some cases, essential for) all types of laboratories. For example, many quality control laboratories do not seek accreditation because the results of their testing activities are not issued to the outside world.

2.3 Proficiency Testing

The aim of proficiency testing is compare a laboratory's performance against that of similar laboratories and thus get a performance measure of its competence. This complements on site assessments and helps the laboratory to monitor its performance and diagnose problems before they become expensive in terms of time and money. In other words, proficiency testing supplements the laboratory's internal quality control by providing external input and complements where utilised, its accreditation by a competent third party.

Proficiency testing can also assist with method evaluation, precision estimation and the availability of reference material for future testing. If the laboratory performs well, proficiency results can also be used as marketing tool which both measure and builds customer confidence.

3 WHAT TYPES OF PROFICIENCY TESTING PROGRAMS ARE THERE?

Proficiency programs fall into two broad categories:

3.1 Interlaboratory Testing Programs

In these programs, a bulk sample of a substance, material or liquid is subdivided. The subdivided samples are then distributed to participating laboratories for concurrent testing. These samples are provided in one of two forms. They may be samples whose properties or measurable components have been set at particular levels (for what is called uniform level experiments). Alternatively, each sample is split into two further sub-samples whose characteristics may differ slightly (called split-level pairs). The laboratory then tests one sample from each sub-level (split-level experiments). This latter case is used where the possibility exists that operators testing replicate samples may be influenced by their results of the previous replicate.

Following the completion of the testing, the results are returned to the co-ordinating body, and analysed against an assigned value (best estimate of the "true value") to give an indication of the performance of the individual laboratories and the group as a whole.

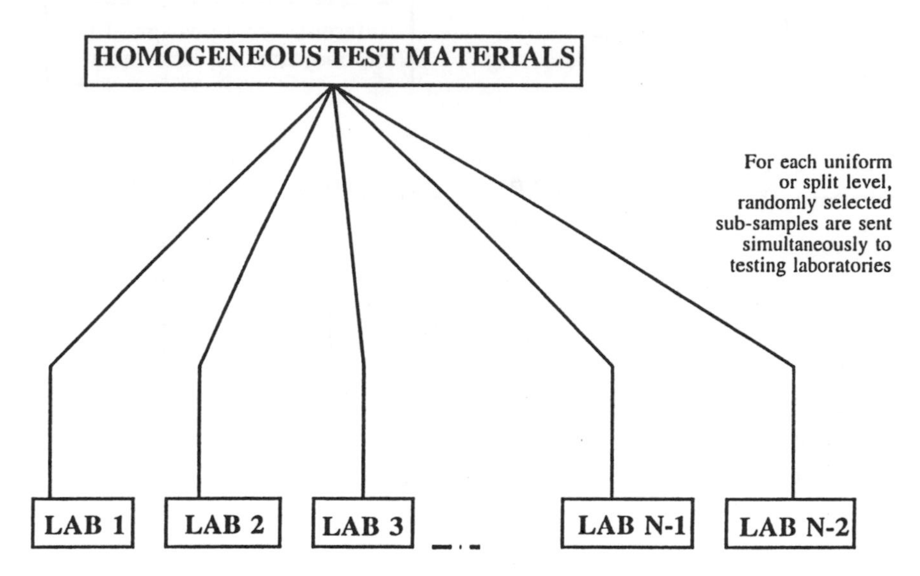

Figure 1 *Typical Interlaboratory Testing Program*

3.2 Measurement Comparison Programs

In these programs, an item (artefact) such as thermometer, resistor or pressure gauge is circulated to laboratories sequentially for measurement of a particular property(ies). The reference value is set at the start of the program by a high echelon laboratory, usually the country's peak authority for the particular measurement under consideration. This laboratory also normally takes responsibility for periodically rechecking the measured properties of the circulated artefact throughout the program to ensure its stability. Programs involving sequential participation take time (in some cases years) to complete. This causes a number of difficulties such as ensuring the stability of the artefact, the strict monitoring of its circulation, the time allowed for testing by individual participants, and the need to supply feedback on individual performance to laboratories during the program rather than waiting till it finishes.

In addition, it may be difficult to compare results on a group basis as there may only be relatively few laboratories whose measurement capabilities closely match each other. Hence, agreement on the values between the participating laboratories, the co-ordinating body and the peak measurement authority must take into account measurement uncertainties and other factors.

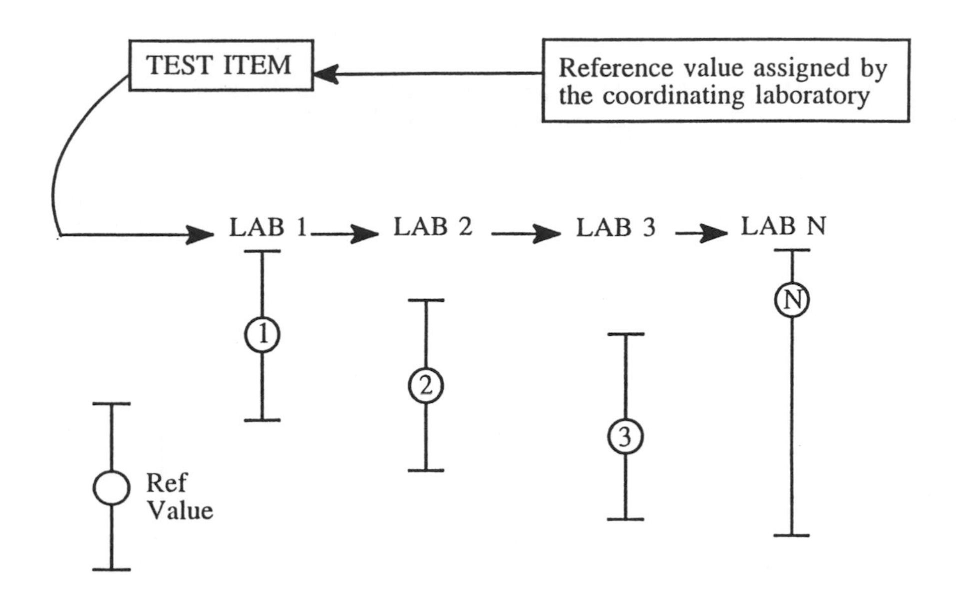

Figure 2 *Typical Measurement Comparison Program*

4 PROGRAM REPORT

Once the results are sent back to the co-ordinating body, they undergo the appropriate statistical evaluation to generate a meaningful and useful appraisal of the program. The presentation of the results is important in conveying the significant aspects of the results at both a group level and for an individual laboratory. The choice of graphical expression of the data will thus depend on the properties or results to be presented. For example, histograms in interlaboratory testing programs allow participants readily to compare their performance with other laboratories. Extreme results are clearly delineated in this type of graph, and histograms will also indicate whether assumptions made earlier about the distribution of the results (e.g. a Gaussian or normal distribution) are valid. Similarly, Youden diagrams are useful where determinations on two similar samples have been performed. These diagrams can highlight systematic problems in laboratories, as well as random errors.

The report on the program is often initially sent to laboratories as an interim report. This provides the laboratory with early feedback about its performance, but may contain only limited statistical information. An interim report is particularly important in protracted measurement comparison programs, where participants cannot wait years to see their evaluation. Summary reports are usually issued at the conclusion of the program and contain more detailed statistical treatment of the results and more complete information on group performance. It may also discuss program design, sample preparation, precision information, and commentary on technical problems, reporting practices, uncertainties of measurement and extreme results.

5 DEALING WITH EXTREMES

Extreme results are those that are inconsistent with, or clearly stand out from, the general spread of results. Care is taken to deal with them early in the analysis so they do not bias the estimation of statistical parameters (e.g. mean, standard deviation). Extreme results alert laboratories to problems with their testing, and thus trigger investigations into the causes.

Common sources of extreme results are incorrect data transfers, inappropriate method use or interpretation, insufficient internal quality control, and staff inexperienced in the method.

It is crucial that the laboratory quickly and thoroughly investigates the cause(s) as other testing areas may also be in trouble. In some cases, the laboratory may be required to report their course of action and their findings to the co-ordinating body, or may seek advice or assistance to determine the source of the problem.

However, it must always be borne in mind that some extreme results may, in fact, be the correct determination, and the statistical analysis should always be applied with this in mind.

6 THE DIVIDENDS OF PARTICIPATION

Many laboratories feel nervous about participating in proficiency programs. However, when properly applied, they are worth their weight in gold as a means of detecting errors and problems that could otherwise remain undiscovered. When the problems are recognised and solved, overall testing performance is improved and both the laboratory and the general community benefit. Proficiency programs also provide reassurance to the laboratory and its clients that procedures, methods and other laboratory operations are under control.

7 CONCLUSION

The combination of laboratory accreditation and participation in proficiency programs and measurement audits is seen as essential as an effective and reliable means of recognising the technical competence of a testing or calibration laboratory. It also provides the foundation for laboratory quality improvement.

International attention is now focusing on the advantages of proficiency testing, particularly with respect to trade where harmonised standards and mutual acceptance of data are paramount if barriers to free movement of goods between countries are to be removed. The time is right for the open exchange of information between operators of programs with the objective of taking a harmonised approach to proficiency testing and promoting good testing and measurement practices. As a result, the number of programs available to testing laboratories worldwide can be increased. This will provide great benefits to laboratories and improve the international traceability and comparability of test results and measurements.

Metrology and the Role of Reference Materials in Validation and Calibration for Traceability of Chemical Measurements

P. De Bièvre,[1] R. Kaarls,[2] S. D. Rasberry,[3] and W. P. Reed[3]

[1] INSTITUTE FOR REFERENCE MATERIALS AND MEASUREMENTS, EUROPEAN COMMISSION JRC, B-2440 GEEL, BELGIUM.

[2] NEDERLANDS MEETINSTITUUT, NL-2600 AR DELFT, THE NETHERLANDS

[3] US NATIONAL INSTITUTE OF STANDARDS AND TECHNOLOGY, GAITHERSBURG, MD 20899, USA

ABSTRACT By the definition of the mole as a base unit for amount-of-substance measures within the International System of Units (SI), chemists can make chemical measurements in full compliance with established metrological principles. Since the mole requires exact knowledge of the chemical entity, which is often neither available nor of practical relevance to the purpose of the measurement, the SI units of mass or length (for volume) are unavoidable in the expression of results of many chemical measurements. Science, technology, and trade depend upon a huge and ever increasing number and variety of chemical determinations to quantify material composition and quality. Thus, international harmonization in the assessments of processes, procedures, and results is highly desirable and clearly cost effective. The authors, with relevant experience and responsibilities in Europe and America, have found some consensus in the interpretation of the metrological principles for chemical measurements, but believe open discussion should precede wide implementation by chemical communities. In fostering this dialogue, this paper shows, for instance, that more precise interpretation of the definitions for "traceability", "calibration", and "validation" is needed for present-day chemical measurements. Problems that face scientists in making measurements do not all vanish just by adherence to the SI. However, such compliance can improve communication among chemists and metrologists.

ACKNOWLEDGMENTS: The authors are very grateful to Steffen Peiser, retired from the U.S. National Institute of Standards and Technology, for his dedicated contributions to the preparation of this publication. They also especially thank two reviewers of the manuscript, Robert Watters and Walter Leight, who provided numerous valuable suggestions.

I. INTRODUCTION:

Science, technology, and commerce require rapidly rising numbers and types of measurements that for good reasons can be trusted [refs.1,2,3,4]. Worldwide acceptance of measurement results requires reliable, traceable, and comparable measurements for reduction of costs, efficient production processes, subsequent use of measurement data, realization of fair-trade conditions, and for internationally recognized and accepted laboratory accreditations. Physical measurements made in accord with the International System of Units (SI), introduced under the Convention of the Meter (with status of an International Treaty), have satisfied many of these needs [refs.5,6]. Such measurements typically rely on a comparison of the measured quantity in the item concerned with the same quantity in a "standard". Chemical measurements are usually not made by comparison with an equivalent chemical "standard". Chemical measurements are not yet widely made in terms of the SI unit of amount of substance, the mole [ref.7]. This paper will explore the possibilities for bringing a stronger metrological foundation to chemical measurements and will specifically describe a role for reference materials in the traceability of chemical measurements to the SI [ref.8].

II. AMOUNT-OF-SUBSTANCE MEASUREMENTS:

Most chemists will agree that the majority of chemical measurements are, or could be, expressed as amount-of-substance measurements. When appropriate, they will in this paper be so described. However, whereas mass or length (volume) measurements at the smallest attainable uncertainty do not generally require a detailed understanding of the material whose property is quantified, amount-of-substance measurements require reference to the exact composition of the measured entity, to interfering impurities, and to the material -- by composition, mass, or volume -- within which that entity is measured.

In many chemical measurements one neither knows nor, at the time of measurement, wishes to know, the exact composition of the matrix. To give an example, a metallurgical firm will receive ore shipments measured by mass in kilograms. Representative samples in the seller's and receiver's laboratories are measured for quality by the amount of substance of a specified metal element or compound per given mass of ore. It is unnecessary and far too complex to attempt amount-of-substance measurements on all components of the bulk. In exactly the same way, a food laboratory might measure the amount of substance (say lead) in orange juice in milligrams per liter (per cubic decimeter). The charm of the SI system lies in a coherence, which makes it possible to express all measured quantities in a combination of base and derived units. [ref.9].

Thus, whereas chemists have historically expressed analyses mostly by mass per mass, or as convenient percentages, or by mass per volume, they could express their measurements in amount of a specific substance per mass (mole per kilogram) or per volume. In cases, such as pure materials and gases, mole per mole can be used. A percentage statement, or one in parts per thousand, million, or billion, is possible, though not recommended. In the SI system, as originally visualized, such dimensionless numbers as results of measurements are not favored. The quantitative result of any measurement should be expressed by a number 'multiplied' by the appropriate unit associated with the measured quantity. As is further discussed in sections IV.6. and IV.12 below, this original preference proposed for the International System does not fit well with much of current practice in chemical measurements.

II.1. Towards Harmony in Amount Measurements: There is no doubt that chemical measurements are and must be widely used in science and research, technology, engineering, agriculture, and in regulatory issues, including boundary crossings, health control, environmental assessment, and commerce. A vast number of chemical measurements are made every year. Ever more will be needed for reason of increasing complexities in human interactions with the environment [refs.4,10,11]. For many measurements worldwide -- such as ozone levels in cities and the upper atmosphere -- it is necessary to maintain anchor points with long-term stability. More generally, all equivalent measurements should be made in harmony with each other [ref.2], even when the practically needed and achievable reproducibility [ref.9] have to be superior to the best attainable uncertainty in measurement relative to "true value". The relation to true value, however, remains the ultimate test for quality of a measurement [ref.12]. At present it is a rather widely accepted opinion that, even when the relation to the true value is elusive, chemists in different laboratories equipped to make repeatable measurements can still make them comparable to one another by the use of a reference material (RM) [refs.13,14,15]. The correctness of this concept will be discussed later (see sections IV 9, IV 10, IV 11, and IV 12).

II.2. The Use of the Mole: We seek to understand the reasons why chemists tend not to express their measurements by the mole, the SI unit of amount of substance, which is said to have been introduced at their request and which is appropriate for many chemical measurements. Some of the background has been discussed previously [refs.7,16]. Here we hope to discuss:

- why and to what extent we advocate a coherent implementation of a wider use of "amount of substance" by chemists;
- why the use of the mole of itself does not solve pressing common problems in chemical measurements;
- why certified reference materials can meet many, but not all, needs of chemists;
- how we hope a consensus either exists or can be achieved regarding the traceability of measurements to SI; and
- why RMs are necessary to promote harmony among chemical measurements worldwide.

II.3. The Nature of Chemical Measurements: Measures of a mass, a length, or a time, are not dependent on the composition and constitution of the material. By the definition of the mole, need exists for amount-of-substance measurements to specify the entity among possibly many types of entities in the material under consideration. Amount-of-substance measurements are highly dependent on the composition and constitution of the material.

Chemical measurements fall into four groups:

II.3.1. Measurements that can be expressed as a mole/mole ratio, the most basic measurements in chemistry, are typified by processes which react, interact, blend, or replace a described amount of substance A with a described amount of substance B. Included are solution concentration measurements when all solutes are known in a known solvent. Note especially that these measurements are independent of the magnitude of the unit mole. Note also that if these measurements are made by mass or volume determinations the uncertainties in the corresponding atomic or molecular mass values must be taken into account.

II.3.2. Measurements that can be expressed as a mole/kilogram or mole/liter ratio are the most commonly made and are typified by a described amount of substance of compound A in an unspecifiable amount of substance B. Note that for these measurements the uncertainty of and relation to the unit mole, just as those applicable to the kilogram or meter, are involved.

II.3.3. Measurements that can only be expressed as kilogram/kilogram or kilogram/liter are unusual because they involve amounts of substances of unknown composition. Instances of this type are not really rare. Examples are particulates in air and condensed-ring compounds in tar. Chemists can be reassured that no mention of the mole is made or needed for expressing the results of such measurements.

II.3.4. Measurements that are described directly in terms of multiples and sub-multiples of the kilogram, the liter, or the mole, are the measurements that provide the underpinning of chemical measurements in science, technology, and trade. They are typified by calibrations or validations of values of weight sets, reference materials, or instruments, as well as by determinations of the magnitude of the unit mole of a specific compound (from the quotient of that compound's mass divided by that of a single ^{12}C atom), or of the Avogadro Constant.

II.4. Measurements for which Reproducibility is More Easily Obtained than Accuracy: Practical chemical measurements are commonly more precise than accurate. By that statement, we mean that the uncertainty of a measurement relative to the true value expressed, in either the mole or the kilogram, is greater than the range for repeated measurements in the same or even different laboratories at different times or by different operators under different environments.

By contrast, satisfactory practical mechanical, electrical, optical, and thermal measurements are often made adequately for the purpose at hand, even if less accurate than corresponds to the optimum achievable uncertainty relative to "true value" expressed under SI. Routine measurements in these fields can thus be expressed conveniently in terms of the relevant SI unit to an uncertainty determined principally by the uncertainty of the practical measurement in the "field"[4]. Harmony among most physical and engineering measurements can be achieved to the uncertainty of the measurement in the field by traceability of all measures to the SI unit without invoking an intermediate "standard"[5] or RM.

In physical science there are occasionally instances where measurements need to be more reproducible than the lowest achievable uncertainty relative to the true value in SI units. Chemists, not just occasionally but as a rule, must achieve traceability of measurements by use of some standard, a reference material, a reference instrument, or a reference method [ref.17]. The spread of these measurements made in different laboratories is often required to be smaller than the uncertainty with respect to true value. Nevertheless, one should state any such measurement in moles along with an assessment of the quality of its reproducibility. Such a statement will be different depending on its applicability within a laboratory, between laboratories, for a given method and environment, or in relation to an RM. When an RM is used, one must also include its often larger uncertainty of traceability to the SI unit. This uncertainty of the value of the RM must be included in the total uncertainty of the unknown.

---

[4] The use of "in the field" is intended without detriment to measurements made in laboratories other than those whose main concern is the traceability link to the true value and the SI.

[5] Although for physical measurements one often speaks of various kinds of "standards" there is a functional difference, but no sharp distinction, in current usage between, say, a transfer standard and an RM.

III. SOME IMPORTANT ISSUES IN ALL METROLOGY:

When discussing traceability of both physical and chemical measurements, one must be clear from the outset on the following conditions applicable to any measurement or measurement capability:

III.1. The Type of Measurement: First we must specify the type of quantity: a base quantity such as temperature (a property that is coupled to a base unit within the SI), or a derived quantity such as pressure (a property coupled to a derived unit, being the quotient of two or more base units within the SI), or even a quantity such as a hydrogen-ion concentration (a property that by convention is not commonly coupled to the SI, although perhaps it should be).

III.2. The Relevant Range of Measurement: Measurements are needed over a total of many more orders of magnitude of a quantity than any one measurement methodology or instrument can achieve. For electric current, the measurement in a nerve fiber near 1 nA will differ from one applicable to a gigantic TA current in a magnet laboratory. At the two ends of the measurement range there is a non-trivial need to relate any "standard" in the smallest or highest range to the applicable SI unit itself. Needed amount-of-substance measurements, too, may range over more than 12 orders of magnitude.

III.3. The Uncertainty Statement: The uncertainty applicable to a measurement contains components for repeatability and reproducibility [refs.9,18,19], caused in part by variability of measurement-relevant influence quantities. The Uncertainty also depends on the individual making the measurement, the laboratory facilities used, and the environment during the measurement. Without some quality control over measurements, statements on relevant traceability can have little meaning. Such controls provide a laboratory with confidence in its operators and credibility to the outside.

Often of general interest is the reproducibility of measurements when operator, equipment or environment are not the same. One must commonly distinguish clearly between uncertainties applicable to measurements at different times (called repeatability [ref.9]) and those made in different places (called reproducibility). A statistical analysis of homogeneity may be needed whenever a measurement is made on a representative sample from the object to be evaluated.

III.4. The Similarity Principle of Metrology: In metrology generally, the closer the similarity between two specimens, the smaller the relative uncertainty of the measured difference between them and the easier it is to make a reliable measurement. Thus, by the use of suitable "standards", measurements in the "field" can become highly reliable and far less demanding and costly.

By this similarity principle it is possible to measure precisely and relatively easily small differences from an amount, or ratio of amounts, given by a "standard". RMs thus become very attractive vehicles for measurement traceability and quality. However, there is an associated problem: good reproducibility of comparisons between pairs of similar specimens is liable to mislead and, in practice, often causes underestimations of total uncertainties through failure to consider additional, large error sources.

III.5. Fitness for Purpose: The achievement of smaller uncertainties than needed is usually uneconomical. In a practical way, realistic uncertainty assessments in relation to true requirements lead to economically sound planning for measurements to be fit for their intended purpose.

III.6. The Classical Measurement Pyramid: A simple view of measurement services pictures the International Bureau of Weights and Measures (BIPM) with its international prototype kilogram and the seemingly perfect constants of physics at the peaks of huge pyramidal systems for all types of measurements, each with many levels. The first level below the apex lists the realizations of units at a number of national metrology institutes, passing on slightly more uncertain measurements to a much larger number of laboratories, which in turn service lower-tier measurement laboratories, until at the very bottom of the broad-based pyramids the workbenches receive calibrations that have become a little more uncertain at every intervening level. That system is simple to understand and works well for most industrial and legal services and for the control of small-scale markets, for which the step-by-step losses from impressive accuracies near the appropriate apex level are tolerable. An inverted pyramid may also become useful for illustrating traceability [ref.7].

For chemical measurements, a possibly preferable system is illustrated in the schematic figure 1. Various possible forms of realization of traceability are given. They range from virtual lack of traceability to a fully "SI-bonded" measurement. The authors tentatively use the term "SI-bonded" to indicate a direct realization of the SI unit, as opposed to being traceably linked by way of measured values. Any user laboratory must seek a reference laboratory that is capable of providing measurement links of the adequate uncertainty and that provides the direct bond to the SI, if that is needed. The reference laboratory can in turn choose the traceability quality that it wishes to maintain, with the responsibility of fulfilling the corresponding competence requirements.

In modern high-technology situations, however, very high reproducibilities are frequently required. A good metrological system must provide means whereby any measurement station can have access to the highest needed level of the system.

IV. SOME IMPORTANT ISSUES FOR THE WIDER INTRODUCTION OF METROLOGICAL CONCEPTS INTO CHEMISTRY:

The above features are common to all measurements. However, some chemical considerations do not have a clear equivalent in physical measurements:

IV.1. The Diversity of Chemical Measurements: Whereas the types of measurements in physics and engineering do not exceed the numbers of base and derived units of the SI, chemical measurements are virtually infinite, equal to the number of chemical elements and compounds. Whereas the magnitude of, say, mass, is defined independent of the entity for which it is measured, the amount-of-substance determination is made specifically relative to one entity. This situation should not lead to confusion, but some chemists fear that it might. For instance the "mole of nitrogen" is not defined until it is said whether reference is made to N_2 or N.

IV.2. The Word "Mole": Other potential difficulties for chemists arise from differences among molecular, molar, and the historic meaning of "mole" in chemistry [ref.16]. Some find the mole unsuited as a base unit in SI because, they argue, it is just a number of entities. Others find the use of "amount of substance" awkward, especially when the entity -- for instance an ion -- is not generally

Fig. 1 : TRACEABILITY SCHEME FOR MEASUREMENT RESULTS

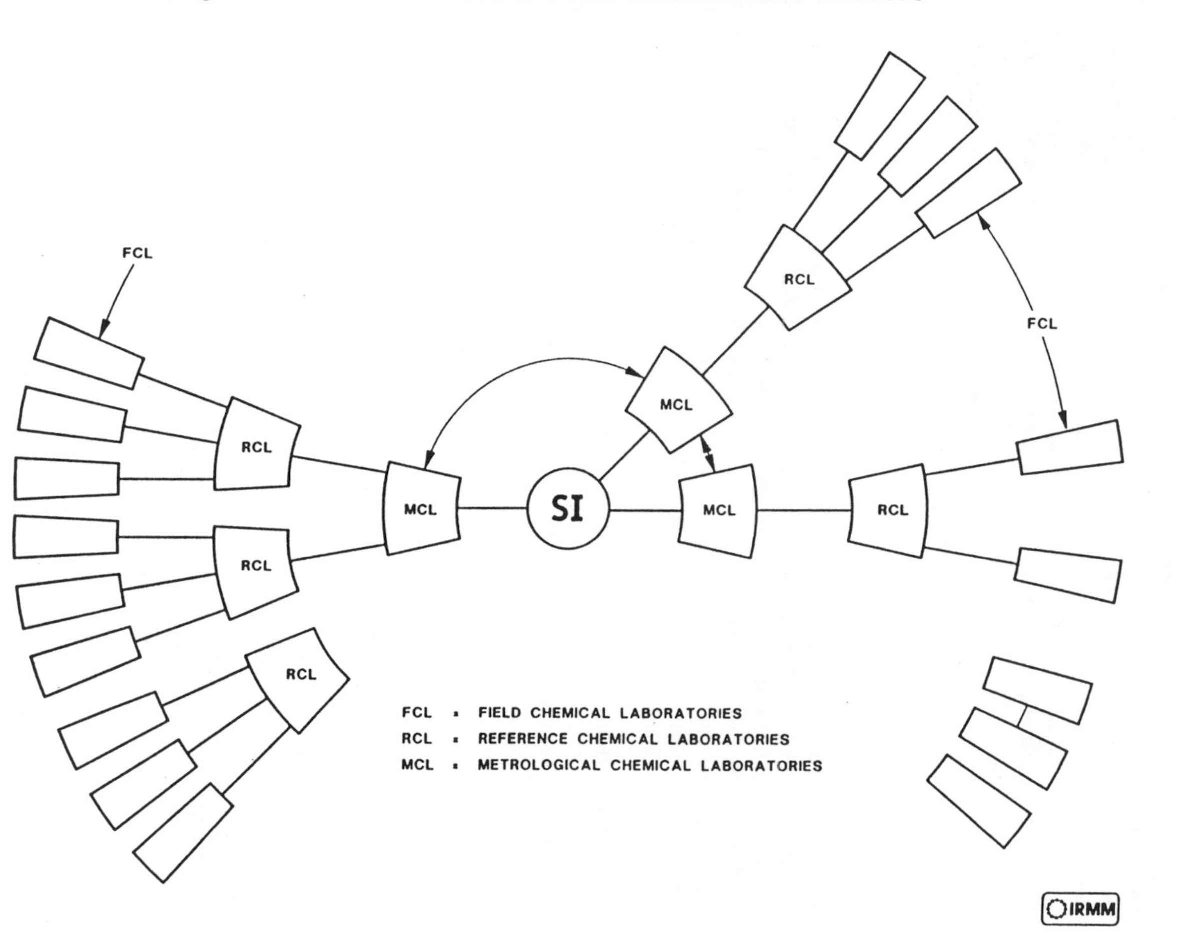

regarded as a substance. However true and relevant are some of these objections to current nomenclature and definitions, a consensus is most unlikely to be reached on any related change in the foreseeable future. Discussion on such a change here is therefore not relevant to more immediate opportunities for a useful consensus in amount-of-substance measurements.

IV.3. The Matrix Effect: Whereas the measurement of, say, mass depends little on the character (*e.g.*, density) of the object for which it is made, that of amount of substance is strongly dependent on the matrix in which the entity resides. Chemists have always been concerned with interferences, but the general problem has become more important with the introduction of many powerful analytical-chemical instruments, the performance of which depends not only on specific physical properties of the entity to be measured, but also on the matrix within which the entity is contained. Chemists may wish for RMs for all entities to be measured in all kinds of matrices of interest to technology or trade. However, the production, of every RM is a time-consuming expensive process. Chemists are thus faced with the unending problem of available resources imposing severe practical limits to the number of RMs that can be produced in conflict with the wide range of matrices of interest. Consequently, a most important contribution that basic chemical science must make is in the development of matrix-independent methods of measurement [ref.20]. The challenge is to separate the one entity to be measured from the influence of all other entities in a mixture. By widespread abilities to do so, metrology in chemistry will reach its most desirable aim to make accurate amount measurements related to the mole unit. In the future, chemical metrology should be directed to the basic science on RMs whose link to SI is strong and on field methods whereby specimens can be compared reliably with the RMs independent of matrix [refs.7,21].

IV.4. SI Recognizes Derived Units (Products or Quotients of Base Units): Whereas the measurement of, say, mass can be stated as a fraction of a total mass (*e.g.*, mass of a sample in a bottle), the amount of substance of a given entity can usually not be stated as a fraction of all amounts of substance in a material. One typically does not even know or care about all the other entities, and one certainly does not generally wish or need to analyze the material in terms of all its constituents. The SI system permits and widely encourages coherently associated units. The substance of interest should, where possible, be expressed in the SI unit, the mole. The other substances, the amounts of which are of no immediate interest to the determination,are quantified in terms of SI units,such as the kilogram, that do not distinguish entities.

IV.5. Do Physicists Use the Mole? Geophysicists generally describe the composition of the universe or of the earth by mass percentages. They could use the mole, the amount of terrestrial substance of, say, aluminum. In the very processes leading to the birth of the elements, amount ratios are of prime interest. The end amount of Al would be expressed in mole per average terrestrial kilogram.

IV.6. Measurement by Ratio: Proponents of the SI for chemistry must consider that proportionality is deeply embedded in chemical thinking[6]. Many of the potentially most reliable analytical techniques -- for instance isotope-dilution mass spectrometry -- yield ratios in the first place. In complex series of ratio measurements the uncertainty propagation is more straightforward than when sums and differences from standards -- such as for mass determinations -- are involved. Consistent with the use

[6] By contrast, many physical measurements are initially additive, as is true for mass, time interval, and length.

of SI, the value of a ratio is called a "measurement" when nominator and denominator are multiplied by a unit and the related uncertainties have been evaluated.

IV.7. Uncertainty about the Nature of the Entity to be Measured: Chemists may not exactly know what is the entity they wish to measure in a material. A common example is moisture, say in grain. There are known to be continuous levels of strengths of chemical bonding of the water molecule in products. Mass loss on heating is routinely used to determine moisture in grain, but may cause error by including in the measured loss volatile compounds other than water, and will also depend on the method used, principally on the temperature and time of drying. In giving the result in mole of H_2O per kilogram, one cannot assure that it was free water in the grain, where some of it was present as a different chemical entity. The same may apply to a metal element, say aluminum in an alloy. The user may well be interested in whether a mole of Al per kilogram refers to total aluminum or just the metallically bound -- as opposed to oxide -- aluminum. The result obtained in a measurement will then depend on the measurement method that is used. The use of amount-of-substance measurements can neither help nor hinder the chemist's need to carefully distinguish significant entity differences such as due to chemical bonding and molecular association in a material.

IV.8. Some Vague Usages of Terms in Measurement Processes [ref.22]: It is quite common for the chemical community nowadays to use the terms "calibrate" and "calibration" for any process that converts an observed value into a more reliable result, which is then called "corrected", "true", or "calibrated". We must also concede that RMs are sometimes used that do not have a matrix closely similar to that of the sample. To make matters worse, uncertainties associated with that situation are generally ignored. Insofar as the chemical community is aware of these problems, the call goes out for more and more RMs in appropriate matrices beyond available capabilities to produce reliable RMs. In order to arrive at rational conclusions on these issues, it is necessary to examine closely and to understand the proper role of "calibration" and "validation" procedures. In the following paragraphs we describe our views and hope that others will endorse them.

IV.9. What Constitutes a Measurement? A measurement of a specified property in an 'unknown' material is a quantitative comparison by ratio or difference made of that property between a reference standard or reference material and the unknown or between relevant settings in an instrument, preferably in the appropriate unit for the quantity under investigation, provided:

a. measurements of the relevant type and range, at the site where the measurement is made, are subjected to reliable uncertainty assessment ;

b. the result (difference or ratio) is proven to be a known function of the true difference or ratio, or appropriately corrected for non-linearity, usually by means of a set of RMs;

c. the comparison applies only to a constituent part of either or both the RM and the 'unknown', and the comparison is:

  i.) proven to be independent of the matrices,

  ii.) based on knowledge that the matrices are precisely similar, or

  iii.) quantitatively evaluated for variability with matrix; and

d. the result is given with its uncertainty including those caused by possible lack of linearity and by the above criteria applied to RMs involved.

Under these conditions, the comparison constitutes a measurement, and the value given of the property in the 'unknown' has been determined.

Chemists will have an important reservation concerning this understanding of what constitutes the uncertainty of a measurement. Physicists and engineers may not, but chemists often are subjected to major sampling, stability, blank, and contamination errors. Chemists should include them in their total uncertainty estimates. The distinction between the measurement uncertainty and the degree to which the measured sample fails to represent the relevant larger bulk needs to be debated and discussed for consensus and understanding.

IV.10. What is a Calibration? Let us begin with the ISO definition [ref.9]: (A calibration is a) "set of operations that establish, under specified conditions, the relationship between values of quantities indicated by a measuring instrument or measuring system, or values represented by a material measure or reference material, and the corresponding values realized by standards". Applied to amount measurements, the "standards" would then be the values assigned to the RMs (of defined composition) at the stated uncertainty relative to the true value of the property, expressed in SI units, or relative to an internationally recognized, certified standard RM for the relevant property, range, and matrix composition.

An instrument or system is said to be calibrated for amount measurements only if within a specified range, a value versus signal (response) curve has been evaluated against RMs including two near the ends of the range. At the present time, it is unfortunately quite common to use the term "calibration" to describe any process which converts a single observed measurement into a more reliable result.

IV.11. What is a "validation"? An RM can validate a measurement procedure (including the measurement instrument) [ref.13] if, prior to its use for an unknown sample, it has been shown to give:

1. a quantitative response for the quantity (in the relevant range) to be measured;

2. a response with a defined and acceptable repeatability;

3. a response with a defined and acceptable reproducibility over changing times and measurement conditions; and thus

4. a defined and acceptable estimate of their overall intrinsic uncertainty.

IV.12. Traceability for Chemical Measurements: The definition by the International Organization for Standardization (ISO) is: "property of the result of a measurement or value of a standard whereby it can be related to stated references, usually national or international standards, through an unbroken chain of comparisons all having stated uncertainties" [ref.9]. Thus, the term does not apply directly to laboratories, but should be applied to the results of chemical amount-of-substance measurements. Every link in the traceability chain should consist of comparisons that are measurements in accord with the above proposed understandings, which include the validation of measurement procedure by RMs. A measurement therefore often has strong links to internationally accepted RMs, but may be only weakly bonded to the SI unit. For comparability among measurement laboratories, the strength of the link must be adequate for assuring equity in trade. Weakness of the bond to SI may thus be acceptable, but the metrologically-minded chemist will ever be disposed to aim for strongly linked reference measurements, methods, and instruments. They are based on simpler concepts with greater permanence, and would be more easily understood by the wider public. Other definitions of traceability have been described [refs.23,24,25].

Not all chemical measurements are, or should be, traceable to the mole. We have seen instances where the unit of mass was the proper SI unit for a quantitative measurement of a material of unspecified entities. There are chemical measurements that are not, but probably should be, referred, and preferably traceable, to the SI unit. Color is used either simply as a qualitative attribute not subject to a measurement, or it is measured quantitatively by some spectrometry, where it may inevitably be subject to high uncertainties from both the measurement itself as well as from theory, such as the Lambert-Beer Law, but well understood in relation to SI.

The description of the relation of a measurement to an SI unit encounters a basic difficulty when the desired meaningful measurement result is a ratio, as in many chemical determinations. The magnitude of the unit then becomes irrelevant. Chemists err when they claim that the inaccuracy of their weight set relative to the international prototype is a component in their uncertainty budget. The self-consistency of their weight set is of course of paramount importance. Since that would include tareweights, internal balance weights, and sensitivity weights, the advice to use weights calibrated against the international kilogram is still generally good.

The quality of ratio measurements seems not to be concerned directly with the SI unit. The only essential condition is that the unit for the nominator be the same as that for the denominator. Traceability requirements for many amount-of-substance measurements, therefore, appear to concern not the unit mole, but a standard measured ratio, preferably between pure defined substances in one RM. Nevertheless, the authors propose that by consensus it shall be a rule for all measurements, where a choice could be made, that it shall fall on the SI unit.

Unusual are measurements for which a direct link to the mole is useful. We should probably not talk about traceability in that connection, because that term is defined as a relation between measured values. An acceptable chain of measurements for compound X of established purity, containing element E that has isotope ^{i}E and that would establish a link to the mole, then would take one of the following general routes: the amount of substance $n(X) \rightarrow n(E) \rightarrow n(^{i}E) \rightarrow n(^{12}C)$; or $n(X) \rightarrow n(E) \rightarrow n(C) \rightarrow n(^{12}C)$. The ratio of atomic masses $m(^{i}E)/m(^{12}C)$ is also involved by the definition, but that ratio is known with a negligible uncertainty compared with the other links in the chain. Clearly only in very few instances will very few laboratories attempt to execute such a chain of measurements for a link to the SI unit. Is it fear that such a difficult process is involved in every chemical analysis that has kept so many chemists from using the mole as the way to express chemical measurement values? Or is it just habit and the convenience of a balance that subconsciously links amount of substance to amount of mass?

IV.13. Laboratory Accreditation: For laboratory accreditation, based on ISO Guide 25 [ref.26,] and the EN 45001 Standard, as well as for certification, based on the ISO 9000 series of standards [ref. 1], it is required that measurement and test results be traceable to international, defined, and accepted physical and physico-chemical standards [ref.27]. This requirement includes the use of the SI. It also includes the proper use of the concept of measurement uncertainty. All these are necessary conditions for reliance on the measurement results of another laboratory. Accreditation is granted when a laboratory has demonstrated that it is competent and capable to work in the above-mentioned sense. Technical trade barriers then fall away, and the needs and requests can be met from industrialists, traders, and the general public in the interest of open and fair trade, health, safety and the environment.

For amount-of-substance measurements we include kilogram mass units, which are linked to the amount-of-substance unit in SI by the atomic-weight values. The latter differ greatly in uncertainty for different chemical entities, but are always available, with the best estimates by current knowledge of their uncertainties, through the International Union of Pure and Applied Chemistry [refs.28,29].

V. REFERENCE MATERIALS:

In the above sections we have already illustrated some of the characteristics and uses of RMs. A more formal definition by ISO is [ref.9]: "material or substance one or more of whose property values are sufficiently homogeneous and well established to be used for the calibration of an apparatus, the assessment of a measurement method, or of assigning values to materials". Extraordinary care in the production of RMs [ref.15] is essential for effective harmonized chemical measurements. Special features of certified RMs are carefully explained by that ISO document [refs.9,15] and their designation as measurement standard specifically authorized.

One may be inclined to suppose that for each type of chemical measurement there is a need to build a measurement system based on the pyramid concept [refs.7,30](section III.6). For the practicing chemist, however, this would be seen only as an unhelpful imposition. Previously discussed limitations of such a pyramid system, would apply equally to the use of RMs. In addition, there is a more major difficulty due to previously discussed differences of RM matrix from that of samples. Whereas for extrinsic measurements, the composition of an RM or other traveling standard is of little or no concern, intrinsic amount-of-substance measurements are generally affected by the internal composition, structure, and texture of the RM.

The limited number of reliable RMs that can be prepared and made available leads to use of possibly inappropriate RMs. When the matrix in a sample differs from that of the RM, reliable comparison may be very difficult. Provision for support of critically important and accurate bench level measurements is needed. In such situations there is a better alternative: From the bench level a specimen with typical matrix properties is sent to a laboratory having competence appropriate to providing a "reference measurement". That value is communicated back to the "bench" where it provides a certified value -- a kind of in-house RM -- for comparison with routine sample measurements. Thus, the concept of reference measurement emerges as equally important as that of the RM. Chemical science has no other choice, since the combined output of RM-producing institutions could not possibly accommodate all the rapidly diversifying demands for all measurands in all matrices of interest.

In order to establish traceabilities of measurements, we advocate the structure shown in figure 1 where many types of linkage can be found including but not limited to those terminating in SI.

V.1. A System for Describing Types of Candidate Chemical Materials for RMs: We would also advocate the optional use of a descriptive materials system for candidate RMs: Firstly, we should have categories depending on the chemical nature of the materials. See Table I, for example.

Table I: Proposed Categories for Reference Materials in Terms of Material Composition

Category	Kind of Material	Description and Criteria in Terms of Material Composition
A	**High Purity**	**Pure Specified Entity** **(Isotope, Element, or Compound) Stoichiometrically and Isotopically Certified as Amount of Substance, with Total Impurities,** **< 10 μmol/mol**
B	**Primary Chemicals**	**As Above, but with Limits of** **< 100 μmol/mol**
C	**Pure**	**One Constituent > 950 mmol/mol**
D	**Matrix**	**Matrix with One or More Major Constituents > 100 mmol/mol**
E	**with Minor Constituents**	**Minor Constituents in Matrix** **< 100 mmol/kg**
F	**with Trace Constituents**	**Trace Constituents** **< 100 μmol/kg**
G	**with Ultra Trace Constituents**	**Ultra Trace Constituents** **< 100 nmol/kg**
H	**Undefined**	**Entities Unspecified or Undefinable**

Secondly, we should agree on RM Classes dependent upon their traceability. See Table II.

The isotopic composition of an element in a specimen can be established and expressed in abundances -- that is amount-of-substance fractions, or moles of isotope per mole of element -- by comparison to synthetic mixtures of enriched isotope Class 0 RMs of that element.

Certification of an elemental Class 0 RM can be performed by metrology laboratories having best scientific procedures under control for the establishment of traceability routes to the SI system. For every such RM the cost in facilities and experts' time is very high and in practice cannot easily be balanced against sales. Only long history of the laboratories' reliability and their free and open discussions of problems coupled with energetic self-criticism will assure the scientific and technological communities. Metrological quality, not cost and economy should be the prime concern of operators within such laboratories.

All other Classes of RMs are needed in much greater number and diversity. They are therefore of much greater potential interest commercially. Intercomparison between similar RMs is always helpful. Only one Class IV or V RM should be made available by consensus for a certain purpose, so that all laboratories are encouraged to make their measurements comparable to others through just one RM.

Table II: Proposed Classes for Reference Materials in Terms of Degree of Traceability to SI

Class	Description and Criteria in Terms of Traceability to SI
0	Pure Specified Entity Certified to SI at the Smallest Achievable Uncertainty
I	Certified by Measurement Against Class 0 RM or SI with Defined Uncertainty by Methods without Measurable Matrix Dependence
II	Verified by Measurement Against Class I or 0 RM with Defined Uncertainty
III	Described Linkage to Class II, I, or 0 RM
IV	Described Linkage Other than to SI
V	No Described Linkage

Validation of a measurement procedure including an instrument can be performed with a Class 0, I, or II RM, but only if differences in matrix or impurities are specified, small, and of proven limited influence on the uncertainty. The uncertainty of the RM relative to true value or the mole may be larger than the link between the measurements on the material and the RM. Traceability between measurements can be achieved with the help of all Classes of RM's, but requires a clear statement on uncertainty. Traceability to the mole, if not by direct realization of the mole, can be established only by Class 0. Their relation to the unit mole must be established by way of atomic-weight determinations or by direct atomic mass comparisons with carbon 12 atoms.

An example of a Class I RM is an RM for which the amount of substance of an element has been measured by isotope dilution against a Class 0 RM, provided the measurement has been shown to be in accord with basic laws [refs.7,21,31] of chemistry and physics.

VI. CONCLUSION:

Reliable chemical measurements in future will depend on more RMs with direct links to the SI, as well as on RMs of greater diversity than are available now. Chemical science will be assisted by clear consensus definitions of traceability, certification, and validation as well as by a widely accepted system for describing RMs by material composition, degree of traceability, uncertainty, quality, and purpose. Ultimately chemists, physicists, and engineers benefit from adherence to the well-grounded and well-established discipline of metrology under a coherent system of units.

VII. REFERENCES.

1. International Organization for Standardization (ISO), *International Standards for Quality Management*, ISO Series 9000, Geneva (1992 and following)

2. B. King, Analyst, 118, 587-591 (1993)

3. S.D. Rasberry, Fresenius J. Anal. Chem.345, 87-89 (1993)

4. P. De Bièvre, Int. J. Environm. Anal. Chem. 52, 1-15 (1993)

5. T. Quinn, IEEE Trans. Instr. Meas.. (In press, 1995)

6. C.H. Page and P. Vigoureux, *The International Bureau of Weights and Measures 1875-1975*, U.S. Department of Commerce, National Bureau of Standards, (1975)

7. P. De Bièvre, Fresenius J. Anal. Chem. 350, 277-283 (1994)

8. International Organization for Standardization (ISO), *Terms and Definitions Used in Connection with Reference Materials*, ISO Guide 30, 2nd Ed. Geneva (1992)

9. International Organization for Standardization (ISO), *International Vocabulary of Basic and* General Terms in Metrology, 2nd Ed. ISO Geneva (1993)

10. P. De Bièvre, Quimica Nova 16(5), 491-498 (1993)

11. A. Lamberty, J. Moody, and P. De Bièvre, Fresenius J. Anal. Chem. in press (1995)

12. P. Cali and W.P. Reed, Natl. Bur. Standards, Spec. Publ. 422, 41-63 (1976)

13. H. Marchandise, Fresenius J. Anal. Chem. 345, 82-86 (1993)

14. W.P. Reed, Fresenius J. Anal. Chem. (In press, 1995)

15. International Organization for Standardization (ISO), *Quality System Guidelines for the Production of Reference Materials*, ISO Guide 34, to be published (1995)

16. P. De Bièvre and H.S. Peiser, Pure & Applied Chem. 64, 1535-1543 (1992)

17. 5th International Symposium on Harmonization of Internal Quality Assurance Schemes for Analytical Laboratories, Washington, D.C., Assoc. Off. Anal. Chem. (1993)

18. International Organization for Standardization (ISO), *Guide to the Expression of Uncertainty in Measurement*, BIPM, IEC, IFCC, IUPAC, IUPAP, OIML and ISO, Geneva (1993)

19. R. Kaarls, Metrologia 17, 73-74 (1981)

20. P. De Bièvre, Ch.7 in H. Günzler, Ed., Akkreditierung und Qualitätssicherung in der Analytischen Chemie, Ch.7. p.131-156, Springer, Heidelberg (1994); Translation into English in Press (1995)

21. P. De Bièvre, Anal. Proc. 30, 328-333 (1993)

22. International Organization for Standardization (ISO), *Management and Quality Assurance Vocabulary*, ISO 8402, Geneva (1986)

23. B.C. Belanger, Standardization News, 8., 22-28 (1980)

24. B.C. Belanger, in 37th Annual ASQC Quality Congress Transactions, p.337-342, American Society for Quality Control (1983)

25. S.D. Rasberry, in 37th Annual ASQC Quality Congress Transactions, p.343-347, American Society for Quality Control (1983)

26. International Organization for Standardization (ISO), General Requirements for the Competence of Calibration and Testing Laboratories, ISO Guide 25, 3rd Ed. Geneva (1990)

27. J.L. Cigler and R.V. White Eds., Natl. Inst. Standards and Techn. Hbk. 150, (1994)

28. Atomic Weight of the Elements, 1993, International Union of Pure and Applied Chemistry, Pure and Applied Chemistry, 66 2433-2444 (1994)

29. J. De Laeter, P. De Bièvre, and H.S. Peiser, Mass Spectrom. Rev. 11, 193-245 (1992)

30. J.K. Taylor, Natl. Inst. Standards and Techn. Spec. Publ. 260-100, 1-102 (1993)

31. P. De Bièvre, Fresenius J. Anal. Chem. 337 766-777 (1990).

Is Total Quality Management a Science?

M. Parkany

ISO CENTRAL SECRETARIAT, GENEVA 20, CASE POSTALE 56, CH-1211, SWITZERLAND

1. INTRODUCTION

The definition of Total Quality Management (TQM) in International Standard ISO 8402 : 1994 is a good starting point for considering what TQM is all about. According to subclause 3.7 :

total quality management

management approach of an ***organization*** *(1.7), centred on* ***quality*** *(2.1), based on the participation of all its members and aiming at long-term success through* ***customer*** *(1.9) satisfaction, and benefits to all members of the* ***organization*** *and to society.*

NOTES

1 The expression "all its members" designates personnel in all departments and at all levels of the ***organizational structure*** *(1.8).*

2 The strong and persistent leadership of top management and the education and training of all members of the ***organization*** *are essential for the success of this approach.*

3 In ***total quality management****, the concept of* ***quality*** *relates to the achievement of all managerial objectives.*

4 The concept "benefits to society" implies, as needed, fulfilment of the ***requirements of society*** *(2.4).*

For further investigation the following paragraphs of the same International Standards may be consulted :

1.7 organization
company, corporation, firm, enterprise or institution, or part thereof, whether incorporated or not, public or private, that has its own functions and administration

NOTE - The above definition is valid for the purposes of quality standards. The term "organization" is defined differently in ISO/IEC Guide 2.

1.8 organizational structure
responsibilities, authorities and relationships, arranged in a pattern, through which an ***organization*** *(1.7) performs its functions*

1.9 customer
recipient of a ***product*** *(1.4) provided by the* ***supplier*** *(1.10)*

NOTES

1 In a contractual situation, the ***customer*** *(1.9) is called the* ***"purchaser"*** *(1.11).*

2 The ***customer*** *may be, for example, the ultimate consumer, user, beneficiary or* ***purchaser****.*

3 The ***customer*** *can be either external or internal to the organization.*

2.4 requirements of society
obligations resulting from laws, regulations, rules, codes, statutes and other considerations

NOTES

1 "Other considerations" include notably protection of the environment, health, ***safety*** *(2.8), security, conservation of energy and natural resources.*

2 All ***requirements of society*** *should be taken into account when defining the* ***requirements for quality*** *(2.3).*

3 ***Requirements of society*** *include jurisdictional and regulatory requirements. These may vary from one jurisdiction to another.*

3.2 quality management
*all activities of the overall management function that determine the **quality policy** (3.1), objectives and responsibilities, and implement them by means such as **quality planning** (3.3), **quality control** (3.4), **quality assurance** (3.5) and **quality improvement** (3.8) within the **quality system** (3.6)*

NOTES

*1 **Quality management** is the responsibility of all levels of management but must be led by top management. Its implementation involves all members of the **organization** (1.7).*

*2 In **quality management**, consideration is given to economic aspects*

2. DISCUSSION

It is evident from the definition that the origin of the concept can be traced back to the **producer-customer** relationship.

All elements of the basic concept have since been greatly broadened.

The producer is now part of the concept of **supplier,** which also includes: distributor, importer, assembler or service organization. The inclusion of service organizations has opened endless possibilities and combinations. **A laboratory is also a service organization,** be it an independent testing house or a medical laboratory of a hospital, a forensic laboratory of the police or an experimental laboratory of a university.

The above examples illustrate also that within an organization there could be parts with special tasks that could also be termed **organizations.**

For example :

A hotel (as an organization) may have a restaurant, a laundry service, shops, a travel agency, a rent-a-car service or a special "hostess service". These can be regarded as organizations within the organization. Further the hotel may belong to an airline company (that may have more hotels) and these all may be owned by a bank.

Similarly, hospitals often have their own kitchen/restaurant, laundry service and choice of funeral parlours. Each of these can be regarded as an organization within a bigger organization.

TQM can indeed be applied in all cases. Because of its relevance now to **society** in which even future generations are included (think of Chernobyl), TQM is an overall, all-encompassing managerial approach that takes into consideration all possible consequences of the activity or inactivity of an organization.

Further examples :

In the past it was quite enough if a carpet for sale was of good quality at a reasonable price. Nowadays we not only have more stringent technical requirements (e.g. fire resistance) but also society has become sensitized and requires proof that the carpets were not made by slave children working in miserable conditions. ISO 8402 subclause 3.7 refers to them when it declares ..."benefits to all members of the organization..", including employees.

One more example : it has been discovered that certain CFC compounds that are used *inter alia* in refrigerators have caused the "ozone hole" that has resulted in increasing occurrences of skin cancer. For the "benefit to society", a new quality requirement had to be added to refrigerator specifications, whereby CFC compounds must be replaced. One can see such refrigerators with "ECO labelling".

TQM is a management approach :

"The strong and persistent leadership of top management and the education and training of all members of the organization are essential for the success of this approach".

We will now examine how TQM can be applied to laboratories. Laboratories, and this should be emphasized, are also "organizations" regardless whether they are independent or parts of a bigger organization, e.g. a laboratory of a factory.

A "one-chemist laboratory" is **not** an organization since "the organization is a managed group of persons and resources formed to carry out a function". In an organization there should be at the minimum several persons in a **structure**.

One chestnut man is not an organization if he does everything alone. (Buying the chestnuts, the charcoal etc. and roasting and selling it). Ten chestnut men doing the same independently is not an organization. However, if they form an enterprise, one being the Director, one responsible for buying the chestnuts and charcoal, another (or a few) doing the roasting, another the delivery, another the selling and another the marketing ... well, then they have indeed formed an organization.

In a laboratory there is a top managing director/chief chemist. Under his leadership there are chemists, laboratory assistants, administrators, cleaning personnel etc. In big laboratories, the number of personnel may be a hundred or more. They may have statisticians, librarians, translators, a purchasing department, first-aid or fire brigade, etc. in one structure.

TQM in such laboratories means that all employees of the laboratory (from the top manager to each member) know their mission, their task and they work in full harmony with each other, with customers/clients to accomplish the task, the mission of the laboratory. Good managers know that employees can be expected to do good quality work if they not only are correctly paid, but also receive appreciation for their achievements, encouragement and support for further education, training, participation at seminars, symposia etc. Appreciation, salary and promotion should be in accordance with merit.

All the laboratory personnel form a team, the members of which communicate and work together with enthusiasm to reach the goals of the laboratory.

The total strengths and abilities of all members of an organization should be fully and effectively utilized and they should be recognized as links in a chain. In an analytical laboratory, the main task is of course to obtain correct analytical results. This depends mainly on the skill of the analyst, who uses validated methods, appropriate and calibrated equipment, certified reference materials, and who regularly participates in interlaboratory studies to increase his proficiency. This is indispensable for customer satisfaction.

The manager of a laboratory is ideally :

- a respected scientist who himself can do any job in the laboratory
- admired by all his subordinates for his humane qualities, his good intentions
- good at communicating with the customers and laboratory personnel alike
- good at organizing the work, with special attention to customer satisfaction and the most economical use of human and material resources
- caring about the laboratory personnel
- caring about the interests of society (environment)

I had the honour of working in a laboratory whose manager had worked much earlier as a young chemist in the team of Professor Szent-Györgyi who received the Nobel prize for vitamin C in 1936.

In his team Szent-Györgyi applied real TQM; although sixty years ago this term was unknown, the concept existed. His assistant (who became later our manager) had learned from him not only biochemistry but also the elements of this concept. We all liked and admired our manager, and we tried to work to perfection, first of all because it was soo good and also because it made him happy.

We had tea together every day for 30-40 minutes and chatted about our work (problems, or how could we do better) and did not know that this was something between "brainstorming" and "benchmarking", as it is called today.

I learned a lot from him concerning the safety of laboratory personnel, and the neutralization and treatment of laboratory effluents and wastes. At that time it was a laudable initiative, nowadays it is a must !

3. SUMMARY

A manager has to win the hearts and minds of his personnel. And TQM is all about this.

I waited until the end to answer the question in the title of my lecture in a **Thesis** ©PARKANY 1995 :

TQM is ***not*** a science, it is more than that:

TQM at its basic level is a ***culture***, at its higher level it is an ***art*** **!**

Method Validation: An Essential Tool in Total Quality Management

J. M. Christensen, J. Kristiansen, Aa. M. Hansen, and J. L. Nielsen

DEPARTMENT OF CHEMISTRY AND BIOCHEMISTRY, NATIONAL INSTITUTE OF OCCUPATIONAL HEALTH, LERSOE PARKALLÉ 105, DK-2100 COPENHAGEN, DENMARK

1 INTRODUCTION

Laboratory errors can be costly as analytical measurements in a modern society serve as a basis for decisions of public authorities for health, safety and environmental protection. Furthermore, the growing interdependence among countries on such social matters illustrates the need for analytical measurements to be comparable.[1] As a consequence, any laboratory may implement systems such as quality assurance (QA) and Total Quality Management (TQM). A vital element is quality assurance describing the overall measures that a laboratory uses to ensure the quality of its operations. This might include a quality system, well trained and educated staff, traceable calibration and documented, validated methods, quality control etc. In this context, an important quality tool is method validation, needed to document the analytical performance and as a tool to reduce measurement errors by correcting results for systematic effects.[2-5]

As the complexity of modern chemical analysis provides many sources of error and opportunities for introducing bias and imprecision, analytical methods ought to be validated *before* implementation. In addition, analytical methods have to be controlled on a regular basis during usage to ensure compliance with the documented performance characteristics or standard procedures.[6] Requirements for harmonization of data require the analytical performance parameters to be documented in the light of a method evaluation study, and the documentation should be related to standards and guidelines issued by international organizations such as ISO, IUPAC, EURACHEM, CEN and WELAC.[7-9]

Several approaches to determine the characteristic key data of a method exist, e.g. comparison with other test methods, intercomparison between laboratories, and an experimental design for method evaluation.[6,10,11] The most frequently used approach for estimating the uncertainty of results is to estimate the imprecision of the method to be used. However, preferably the analytical performance should be expressed by several test parameters, e.g. range, linearity, accuracy, limit of detection (LOD), limit of quantification (LOQ), specificity, and ruggedness as described in standards or guidelines issued by international organizations such as ISO, IUPAC, CEN.[4,8] A recently published design of method evaluation evaluates the performance characteristics for the analytical method by estimating the method evaluation function (MEF) resulting in an estimation of the total errors of the analytical method, including the within- and between-run variability and the systematic effects. The MEF design is based on a linear least squares regression analysis of the measured concentration vs. the conventional true concentration of a series of method evaluation samples containing the analyte in the expected linear range of the method. Statistical tests for linearity of the regression line and normal distribution of the residuals (N-Score test) are performed. The method evaluation ensures that the experimental standard deviation is a valid measure of uncertainty contributions from *random effects* and *systematic effects* such as contributions from different analysts, calibration uncertainty, scale graduation errors, equipment and laboratory. Furthermore, the design ensures that the basic assumptions for statistical tests for systematic effects are fulfilled. The systematic effects can be quantified as zero point errors, if the intercept is significantly deviating from zero, and proportional errors, if the slope is significantly deviating from unity.[11]

As test materials reference materials (RM) are finding increasingly use, and when using certified RM for method validation they are invaluable tools for providing traceability. When possible, the analysis of several certified reference materials covering the concentration range of interest is the most useful way to investigate measurement bias; however such documentation is still lacking in many studies. When the method evaluation study demonstrates significant systematic effects, such errors should be corrected using the MEF equation. Based on two specific examples demonstrating the needs for correcting measurement bias in analytical chemistry the present paper discuss the importance of method validation as an essential quality tool in TQM.

2 EXPERIMENTAL

In a biomonitoring program on measurement of hormones, adrenaline, noradrenaline and prolactin were selected as biomarkers for work related stress. Adrenaline and noradrenaline were measured by high pressure liquid chromatography (HPLC) and prolactin by immunoassay. The analytical methods were validated before used.

2.1 Sampling of blood and urine. Blood samples were taken by venepuncture in Vacutainer® tubes. After centrifugation the serum was stored at -20°C until analysis. Urine spot samples from white-collar and blue-collar workers were collected in 50 ml polypropylene bottles containing 1.0 g citric acid and the urine samples were stored at -20°C until analysis.

2.2 Sample preparation for the method evaluation experiment. A urine pool was collected from environmentally exposed subjects at low stress exposure, to be used for the preparation of a batch of samples for method evaluation (ME-samples). A batch of ME-samples was prepared by dilution of adrenaline and noradrenaline stock solution in urine in the expected linear range of the method (3 - 180 nmol dm^{-3} for adrenaline and 15 - 350 nmol dm^{-3} for noradrenaline). All ME-samples were prepared in duplicate, and were analyzed in random order. Only two concentration levels were determined in each run, and each ME-sample was determined twice on different days. All ME-samples were treated as real samples for analysis.

2.3 HPLC analysis of adrenaline and noradrenaline in urine. A fast HPLC method for quantification of the catecholamines adrenaline and noradrenaline was developed for a project on work related stress.[13] Epinephrine (99%), and norepinephrine (99%) for the production of calibrants were obtained from Aldrich Chemical Company Inc., Milwaukee, USA. Five different calibration solutions containing adrenaline and noradrenaline within the concentration range 0 - 180 nmol dm^{-3} and 0 - 350 nmol dm^{-3}, respectively, were made by spiking a pool of human urine obtained from white-collar workers.

The urine samples were analysed by HPLC without any pretreatment. The HPLC system included an isocratic and an intelligent Merck Hitachi HPLC-pump series L 6200A, a variable fluorescence detector model LS-4 (Perkin Elmer), a WISP 710B auto-sampler and Millenium chromatography software from Waters Associates. The samples were purified automatically on a biogel Probenkoncentrator column and separated on a LichroCart column packed with RP C-18 (5 μm particles). Unfortunately, no certified reference materials are available for traceability of the analysis of adrenaline and noradrenaline in urine. Therefore, for quality control a batch of samples was produced by spiking a pool of urine with adrenaline and noradrenaline, obtaining a concentration of 50, 150 and 100, 300 nmol dm^{-3}, respectively. The between-run imprecision expressed as standard deviation was 0.97 and 1.17 nmol dm^{-3} in the range 0 - 180 and 0 - 350 nmol dm^{-3} and the limit of detection was 2.90 and 3.51 nmol dm^{-3}, respectively[13].

2.4 Immunoassay for analysis of prolactin in serum. Assays for determination of prolactin in serum were a double-antibody ^{125}I radioimmunoassay, RIA (Diagnostic Products Corporation, Los Angeles, USA) and a time-resolved fluoroimmunoassay, Delfia® (Wallac, Turku, Finland).

In the RIA procedure, ^{125}I-labelled prolactin spiked to the sample competes with prolactin for sites on prolactin-specific antibodies. After incubation, the separation of bound from free prolactin was achieved by the PEG-accelerated double-antibody method. The tube was counted in a gamma counter (1470 Wizard™, Wallac) equipped with MultiCalc PC software (Wallac).

In the DELFIA® one-incubation procedure, samples were reacted with europium-labeled antibodies and with immobilized monoclonal antibodies directed against a specific antigen site on the human prolactin molecule. Enhancement solution dissociates europium ions from the labelled antibody into solution where they form fluorescent chelates with components of the enhancement solution. The fluorescence was measured in a time-resolved fluorometer (1234 Delfia® Fluorometer, Wallac) connected to MultiCalc PC software (Wallac). The inter-assay (between run) and intra-assay (within run) imprecision of results and the limit of detection for the assays is presented in Table 1.

2.5 Statistics. Method evaluation was carried out as described previously.[12] Any chemical method can be characterized by its method evaluation function (MEF), defined as the functional relationship between a conventional true value of the analyte (μ) and the estimated result of the chemical analysis (μ_Y). The equation of MEF is: $\mu_Y = \alpha + \beta\mu$ characterized by its intercept (α) and slope (β). The method evaluation is performed by

Table 1 Performance characteristics for the double-antibody RIA-assay and the DELFIA-assay for quantitative measurement of prolactin in human serum (mIU dm^{-3}).

Performance parameter[a]	RIA	DELFIA®
Inter-assay (between run):		
SD	12.0	8.4
CV %	8.3	4.7
Intra-assay (within run):		
SD	5.8	3.1
CV %	4.0	1.7
LOD[b]	20	9

[a] in the range 150-200 mIU dm^{-3}.
[b] 3 standard deviations above the mean of a "low" measurement value.

statistical analysis of the MEF. The underlying theory is based on the assumption that the analytical method is in statistical control, i.e. the variation among the observed results can be attributed to a constant system of chance causes and approximated by a normal distribution.

Test for normality and linearity is performed.[15] When the assumption is correct, the standard deviation ($\sigma_{Y|\mu}$) of Y given μ is an expression of the variation of the method [4]. The systematic error consists of two components, the zero point (α) error and the proportional error (β-1). Ideally, an analytical method should be without any systematic error, i.e. $\alpha = 0$ and $\beta = 1$. However, in practice it is not possible to ensure $\alpha = 0$ and $\beta = 1$ for all values of μ. When a least-squares regression analysis of the MEF is performed and the standard deviation of α and β is calculated, the accuracy of measurements is validated by testing if $\alpha = 0$ and $\beta = 1$ using a t-test.[12]

Since linear least-squares regression analysis is based on the assumption of constant standard deviation, weighted regression analysis needs to be performed to estimate α, β and $\sigma_{Y|\mu}$, if the variation of the method increases with increasing conventional true concentration.[16] The square root of the relative mean square error (RMSE^{½}) is used for the simultaneous estimate of the random and systematic effects:

$$RMSE^{1/2} = [[(\alpha/\mu + (\beta - 1)]^2 + (\sigma_{Y|\mu}/\mu)^2]^{1/2} \quad (1)$$

Manual implementation of method evaluation is difficult and therefore the AMIQAS PC-quality assurance program for DOS and Windows, meeting the demands in ISO standards and guidelines, was used for statistical evaluation.[16]

The statistical analysis for evaluation of the prolactin assays was based on matched pairs with errors in both variables (different standard deviation). The group of paired data consists of the measurements on the same serum samples measured by two different methods, i.e. DELFIA-assay and RIA-assay. The statistical model is a linear functional relationship between the mean values of individual measurements:

$$\mu_1 = a + b\mu_2 \quad (2)$$

where μ_{1i} and μ_{2i} denote the mean values of the DELFIA and RIA measurement on serum sample no. i, a is the intercept and b the slope estimated by the model. It is assumed that the values of a and b are the same for all serum samples. The purpose of using the model (2) is that this model is not assuming the same change in μ_1 for all values of μ_2. The estimation of the parameters a and b was as described by Mandel.[17]

3 RESULTS AND DISCUSSION

The psychosocial working environment is an area with an increasing attention to the effort of developing a healthy working environment. In this context, measurement of biomarkers for stress is attractive as an objective measure of work related stress.

Several biomarkers have been used in connection with work related stress, but catecholamines are most frequently used and thoroughly examined in connection with other stress influences than work. Levels of catecholamines in serum and urine change very quickly through a stress exposure. Adrenaline has turned out to be more related to mental influence while noradrenaline is related to vascular and muscular contractions. The biomarker prolactin in serum has turned out to correlate with passiveness in crisis situations and it is expected that prolactin is suitable as a biomarker for personally experienced situations of stress in the working environment.[18]

Biomarkers for stress provides the opportunity of identifying subjects and subgroups influenced by mental and physical pressure, thereby recognizing possible health risks due to work-related stress. In such biomonitoring programs it is necessary to follow the same individuals for years, e.g. before and after an intervention of their job function. It is of the utmost importance that obtained measurement results are comparable, and therefore the analytical methods used in biomonitoring programs must be validated and corrections for measurement bias carried out.[19] Unfortunately, it is still unusual that a measurement bias demonstrated in a method evaluation study is used to correct measurement data, although it is required in recently issued international guidelines.[8,20-22]

For the present study, the method evaluation parameters for adrenaline and noradrenaline are given in Table 2. The corresponding method evaluation plot, i.e. the measured concentrations plotted against a conventional true value of adrenaline in urine, is shown in Figure 1. Since uncertainty increases with increasing concentration, weighted regression analysis was used to estimate the MEF. The result of the N-score test shown in Table 2 revealed that the distribution of the MEF points around the line of regression did not deviate significantly from normal distribution for noradrenaline. However, for adrenaline the N-Score test indicated that the method was not in statistical control and, therefore, it has to be improved before any parameters are used for correction.

The pure error lack of fit test for linearity demonstrated that the MEF for noradrenaline did not significantly deviate from linearity, and the test for systematic errors (α and β) showed that the MEF had no zero point error, i.e. α was not significantly different from zero (Table 2). However, the method evaluation study demonstrated a significant systematic proportional error for noradrenaline in urine ($\beta = 0.84$) more distinctly exposed in a RMSE^{½} plot.[12] It is important to emphasize that the number of ME-samples included in a method evaluation study should be sufficiently high to produce an acceptable SD of the slope (β) to be used for correction of results, i.e. SD(β) < 0.025 is able to expose a proportional error $|\beta-1| \geq 0.05$.[12] In the present study, the SD(β) was 0.02 for noradrenaline; therefore, the number of ME-samples has been sufficient to obtain an acceptable MEF slope to be used for correction of bias. Moreover, in a comparison study with a new RIA-assay from IBL (Hamburg, Germany) it was demonstrated that the

Table 2 *Method evaluation key parameters for the catecholamines adrenaline and noradrenaline in urine (nmol dm^{-3}).*

Performance parameter	Adrenaline	Noradrenaline
Slope (β)	0.92*	0.84*
SD (β)	0.02	0.02
Intercept (α)	-1.81**	0.40
SD (α)	0.48	0.82
Pure test	0.15[a]	0.40[b]
N-Score	0.86[c]	0.99[c]
$\sigma_{Y\|\mu}$ [d]	0.97	1.17
LOD[e]	2.90	3.51

* Significantly different from 1, $p < 0.05$; ** Significantly different from 0, $p < 0.05$
[a] Table value: 2.68; [b] 3.05; [c] 0.95
[d] Estimated standard deviation on the method evaluation function
[e] Limit of determination (LOD) defined as the concentration resulting in RMSE^{½} = 33%.

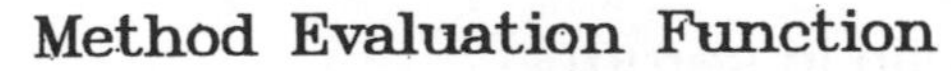

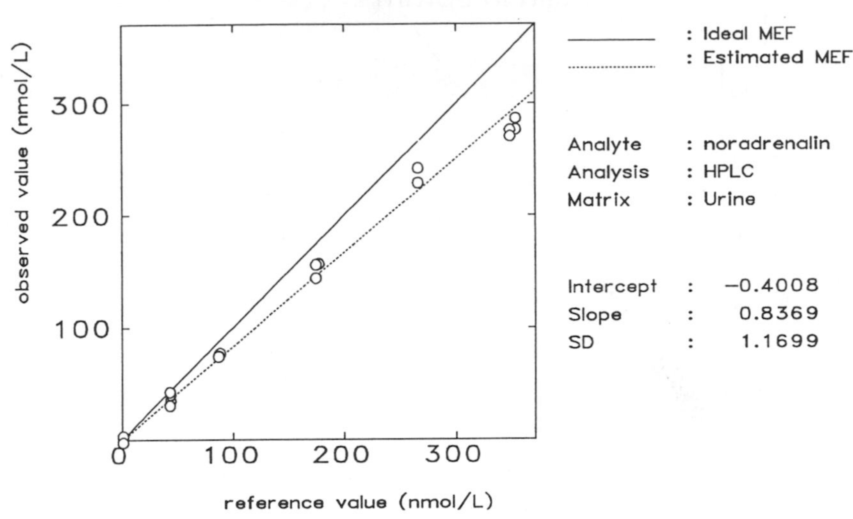

Figure 1 *Method evaluation plot of a method evaluation study on noradrenaline in urine measured by HPLC.*

HPLC method as well as the RIA method were biased. Consequently, when using the HPLC method for measurement of noradrenaline in urine, the systematic error was corrected by dividing the measured values by 0.84. This example illustrates the necessity of a thorough method evaluation study before measurement errors are corrected using key parameters in a method evaluation study.

For evaluation of the prolactin assays a method evaluation study was considered, and for the preparation of the ME-samples a standard material was required. Within the last decade there has been a need for an international standard (IS) for human prolactin since immunoassay is extensively used in clinical medicine, and the purified hormone is scarce, heterogeneous and unstable and consequently difficult to characterize by bioassay procedures. As a consequence, the first IS was established in 1986 and a the Second International Standard for prolactin was established in 1988 (IS coded 84/1500) with a value of 53 mIU activity/ampoule of prolactin assigned by the expert Committee on Biological Standardization of the World Health Organization (ECBS).[23] Unfortunately, attempts to dissolve the lyophilized material in any biological matrix were not successful. Although it was possible to dissolve the material in alcohol, the alcohol could influence the protein matrix and is, therefore, not suitable in connection with immunoassay. Consequently, to evaluate the relationship between the RIA-assay and the DELFIA-assay real samples were measured in both assays and a linear relationship was estimated using a functional model as shown in Figure 2.[17] The functional relationship between the individual mean values was estimated ($\mu_{DELFIA} = 17.32 + 1.2287\mu_{RIA}$). The information drawn from this relationship was that using the DELFIA-assay the obtained measurement results were considerably higher than the results obtained by the RIA-assay. At least one of the methods has a bias, but it is impossible to conclude which method is biased due to lack of standard material.

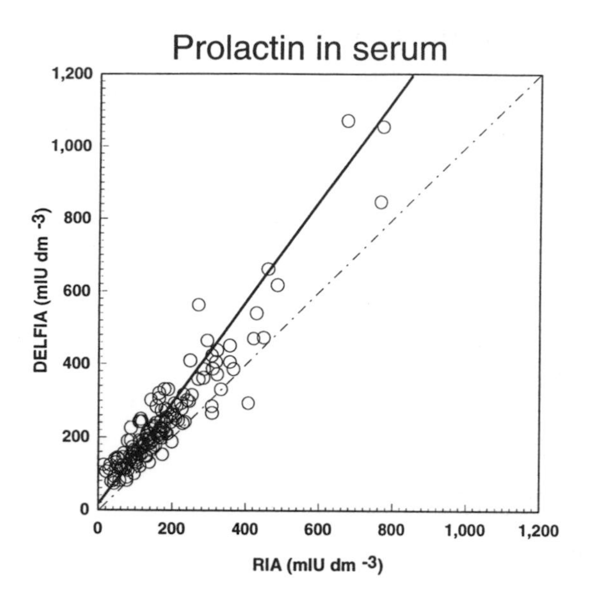

Figure 2 *Results of prolactin analysis performed by RIA-assay and DELFIA-assay on the same samples. The functional model between the mean values of the same sample measured using RIA-assay and DELFIA-assay is given by the solid line. The dotted line is given as the corresponding identical line.*

It has been emphasized that the influences of the matrix on measurement results depend on the type of procedure, equipment etc.[5] An analytical bias may, therefore, be detected by comparison of results for the same samples obtained by different analytical methods, preferably relying on different analytical principles. In the absence of any analytical bias, the functional relationship between the results obtained by the DELFIA and RIA methods applied to the same samples is a straight line from origo and a slope equal to unity. It is unlikely to believe that the method has the same bias. However, the RIA and DELFIA assays are based on fairly identical analytical principle (immunoassay); thus it is impossible to appoint the true values. This illustrates the need for certified reference material of prolactin in serum to make correction for systematic effects. The functional relationship established in this study may be used to establish harmonized results, i.e if a change of measurement procedure is accomplished a correction has been taken into consideration. Alternatively, a calibration could be carried out by collaboration studies, i.e. participation in proficiency testing schemes or external quality control schemes (EQAS), where target values are established by reference laboratories. Samples obtained from these studies are very suitable to be used in a method evaluation study, if stable over a period of time.[2]

When an analytical method is developed by a laboratory or a standard method is used for routine purposes, it is required to evaluate the key characteristics such as accuracy, repeatability, reproducibility, sensitivity, limit of detection, selectivity etc.[4,8] In a method evaluation study of valproate in plasma measured by SYVA's EMIT-assay, the method evaluation study evidently demonstrated a systematic proportional error of the EMIT-assay (β = 0.95). Measurement results obtained by this method were corrected by dividing the measured values by 0.95 and the key characteristics determined.[2] Metal analysis is another important area for which method evaluation has been applied and measurement error corrected.[6,12,19]

According to the recently issued ISO guide on expression of uncertainty, the uncertainty components of the method are combined to a total uncertainty as "an estimate attached to a measurement which characterizes the range of values within which the true value is asserted to lie", at a probability of 0.95.[9,20,21] A statement of the uncertainty associated with the results persuades the "user" about the quality of the result. Therefore, the presented method evaluation designs for validating analytical procedures are prerequisites for documentation of the key parameters. Moreover, validation also includes the ability to understand the needs of the user, the ability to select proper measurement methods to serve the needs, especially the uncertainty and traceability.[4,8]

A considerable literature already exists on method validation, but it has to be stressed that the validation of the method should cover all those steps that influence the final result of measurement, e.g. sampling, transport of the samples, preparation of samples, the various analytical steps, the quality of the reference materials and factors due to evaluation of raw data and calculation of the final results. [19,24,25]

In a TQM system method validation is an integrated part of a quality system and an important tool. TQM concerns optimum utilization of resources, among them the human resources, by education, training, and acknowledgement of the fact that the individual feels satisfaction in delivering high quality data. Although it takes time to implement new concepts such as method validation, the outcome will be a highly motivated staff. This is the most effective method to secure valid and reliable data that can be used and interpreted by others than the producer of the results. Standardization alone will not result in traceable and comparable measurement results, but a combination of standard regulations for method validation and initiatives for implementation of the new uncertainty concept is necessary measures to achieve harmonized data.

References

1. B. King and R. Walker, *Anal. Chem.,* 1994, **66**, 1168A.
2. J.M Christensen, *Mikrochim. Acta.*, Submitted 1995.
3. ISO 8402. Quality management and quality assurance vocabulary. International Organization for Standardization, Geneva, Switzerland, 1994.
4. D. Holcombe. Quality Assurance for Chemical Laboratories. CITAC Guide CG1, 1st draft Jan 1994.
5. K. Heydorn, *Mikrochim. Acta.*, 1991, **III**, 1.
6. J. M. Christensen, O. M Poulsen, T. Anglov, H. G. Seiler, A Sigel, H. Sigel (eds) 'Handbook on metals in clinical and analytical chemistry', Marcel Dekker, New York, 1994, Chapter 4, p. 45.
7. EN 45001. General criteria for the operation of testing laboratories. Brussels: CEN 1991.
8. WELAC/EURACHEM. Accreditation for chemical laboratories: Guidance on interpretation of the EN 45000 series of standards and ISO/IEC guide 25. Guidance document WGD2/No1, 1993.
9. Guide to the expression of uncertainty in measurement, 1st edition. International

Organization for Standardization, Geneva, Switzerland, 1993.

10. W. Horwitz, *Pure & Appl. Chem.*, 1990, **62**, 1193.
11. W. Horwitz, *Pure & Appl. Chem.,* 1988, **60**, 855.
12. J. M. Christensen, O. P. Poulsen, T. Anglov, *J. Anal. Atom. Spectrom.*, 1992, **7**, 329.
13. Aa. H. Hansen, J. L. Nielsen, J. M. Christensen. In preparation 1995
14. J. C. Miller and J. N. Miller, Statistics for analytical chemistry, Chichester: Ellis Horwood Limited, 1993. Ed. 3.
15. P. Armitage and G. Berry. Statistical Methods in Medical Research 2nd Edition, Blackwell Scientific Publication, Oxford, 1971; 266-270.
16. S. L. Christensen, J. T. B. Anglov, J. M. Christensen, E. Olsen, O. M. Poulsen, *Fresenius J. Anal. Chem.,* 1993, **345**, 343.
17. J. Mandel, *J. Qual. Tech.*, 1984, **16**, 1.
18. AA. M Hansen, Biomarkers for Work Related Stress, Report, The Danish Working Environment Fund, Copenhagen, Denmark, 1995.
19. J. M. Christensen, Human Exposure to Toxic Metals. Factors influencing interpretation of biomonitoring results, *Science Tot. Environ.*, In Press 1995.
20. The Expression of Uncertainty in Testing, WELAC,. Draft document WG 5. 1994.
21. EURACHEM, Quantifying uncertainty in analytical measurements, Draft version 5, Eurachem. September 1994.
22. ISO/REMCO N 319. Traceability and calibration in analytical chemistry and material testing; Principles and applications to real life, in connection with ISO 9000, EN 45000 and ISO Guide 25, June 1994.
23. D. Schulster, R. E. Gaines Das, S. L. Jeffcoate, *J. Endocrin.*, 1989, **121**, 157.
24. J. Kristiansen, J. M. Christensen, J. L. Nielsen, *Mikrochim. Acta*, Submitted 1995.
25. Draft ISO/DIS 9004-5. Quality management and quality systems- Part 5: Guidelines for quality plans. International Organization for Standardization, Geneva, Switzerland, 1994.

How to Implement the ISO/BIPM/OIML/IUPAC Guide to the Expression of Uncertainty of Measurement in TQM

Jesper Kristiansen,[1] Jytte Molin Christensen,[1] and Kaj Heydorn[2]

[1] NATIONAL INSTITUTE OF OCCUPATIONAL HEALTH, DK-2100 COPENHAGEN, DENMARK

[2] RISØ NATIONAL LABORATORY, DK-4000 ROSKILDE, DENMARK

1 INTRODUCTION

Reliable and comparable measurement data are important for decision-making in a modern society. A measurement result is therefore useless without an estimate of its uncertainty, which is a quantitative measure of the reliability of the result. Recently, the *Guide to Expression of Measurement Uncertainty*[1] was published jointly by ISO, IUPAC and other organizations in order to harmonize the expression and evaluation of uncertainty in measurement. The BIPM approach for estimating uncertainty, presented in this guide, breaks with the traditional approach based on statistical evaluation of repeated measurements.[2] The two approaches differ significantly and may not yield identical estimates of the uncertainty. Distinct features make the BIPM approach preferable over the traditional approach: Firstly, it allows statistical control of the measurement process to be examined and documented, secondly, it provides a tool for optimizing the measurement process by identifying critical steps, and thirdly, it is the only approach that takes into account the traceability of measurement results to primary standards. There is an increasing demand from society, *e.g.* from accreditation bodies, that measurement uncertainty is evaluated according to the BIPM approach, but in spite of this fact, and in spite of the obvious advantages of the BIPM philosophy, the guide[1] is far from being widely accepted by the analytical chemistry community. Based on a concrete example from analytical chemistry the present paper discuss the advantages of the BIPM approach to measurement uncertainty.

2 REPEATABILITY, REPRODUCIBILITY AND UNCERTAINTY

The uncertainty quantify the "doubt about the exactness of a measurement result".[1] Before considering a specific example on evaluation of measurement uncertainty, it will be worthwhile to compare the concept of uncertainty with other measures of "doubt about the exactness measurement results", specifically repeatability and reproducibility.

Repeatability and reproducibility are measures of precision, which is defined as the closeness of agreement between independent measurement results obtained under stipulated conditions.[3] It should be noted that this definition connects repeatability and reproducibility with observations, *i.e.* estimates of repeatability and reproducibility are based on *a posteriori* information.[4] Obviously, the number and type of experimental conditions that are varied while the replicate measurement results are obtained will influence the observed variability among the results (Figure 1). Therefore, all the conditions that have been varied must be stated together with the number of measurements when values of repeatability and reproducibility are reported. Clearly, this property of repeatability and reproducibility makes demands of the honesty and integrity of the analyst who estimates these parameters. Even if the conditions for estimating the standard deviations are reported, different measurement results are only directly comparable if the same experimental conditions have been varied.

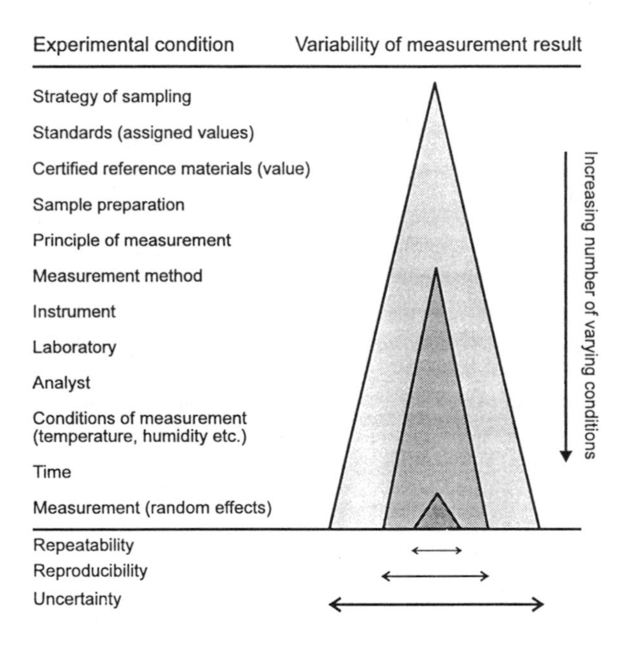

Figure 1 *Variability of replicate measurement results originates from non-constant conditions of measurement*

Unlike repeatability and reproducibility, the definition of uncertainty does not presuppose the obtainment of measurement results. According to ISO standard 3534 uncertainty is an estimate attached to a measurement result which characterizes the range of values within which the true value is asserted to lie,[3] which indicates that uncertainty is a measure of accuracy,[3] which includes both precision and trueness. In principle, the uncertainty of a measurement result can be obtained experimentally - like repeatability and reproducibility - by varying the experimental conditions. However, in practice a measurement result depends on more parameters than it is possible to vary in an experiment (Figure 1),[1] and therefore the uncertainty has to be evaluated using a mathematical model to describe the relation between the parameters and the measurement result. Since uncertainty can be evaluated prior to any measurements, it is said to be based on *a priori* information.[4] In general, if the dependence between a measurement result X and the parameters Y_1, Y_2, .., Y_i, .., Y_n are given by the function G:

$$X = G(Y_1, Y_2, \ldots, Y_i, \ldots, Y_n) \tag{1}$$

then an approximate relation between the variance of X, σ_X^2, and the variance of the parameters Y_i, σ_{Yi}^2, is given by the expression:

$$\sigma_X^2 = \sum_{i=1}^{n} \left(\frac{\partial X}{\partial Y_i}\right)^2 \sigma_{Y_i}^2 \tag{2}$$

for uncorrelated parameters.

3 UNCERTAINTY IN THE DETERMINATION OF LEAD BY AAS

Lead in human whole blood can be determined by atomic absorption spectrometry (AAS). In principle, a blood sample is heated to bring the atoms up in a gas phase. In the gas phase the lead atoms absorb light at a specific wavelength, and the absorbance is recorded. The signal is the time-integrated absorbance. To convert a sample signal to concentration a calibration curve is constructed from standards with known concentrations.

The mathematical model for describing the measurement process is:

$$C = f \cdot \frac{Y - Y_0}{b} \tag{3}$$

where C is the concentration, Y and Y_0 are respectively sample and blank signal and b is the slope of the calibration function. A correction factor f has been introduced to compensate for a systematic effect on the measurement result from gradual wear of the atomization site during measurement. An approximate relation between the uncertainties of the parameters on the right side of Eqn. 1 and the uncertainty of C is given by:

$$\sigma_C^2 = \left(\frac{\partial C}{\partial f}\right)^2 \sigma_f^2 + \left(\frac{\partial C}{\partial Y}\right)^2 \sigma_Y^2 + \left(\frac{\partial C}{\partial Y_0}\right)^2 \sigma_{Y_0}^2 + \left(\frac{\partial C}{\partial b}\right)^2 \sigma_b^2 \tag{4A}$$

or, after obtaining the partial derivatives and inserting in the expression:

$$\sigma_C^2 = C^2 \left[\left(\frac{\sigma_f}{f}\right)^2 + \left(\frac{\sigma_b}{b}\right)^2\right] + \left(\frac{f}{b}\right)^2 (\sigma_Y^2 + \sigma_{Y_0}^2) \tag{4B}$$

In deriving the Eqn. 2 the absence of correlation between f, Y, Y_0 and b was assumed.

The standard deviation of the sample signal, σ_Y, has contributions from several uncertainty components: Prior to measurement the sample is diluted and pipetted to the graphite furnace of the spectrometer for measurement; therefore, σ_Y is proportional to the relative uncertainties of the dilution factor and the pipetted volume. In addition to these components there will be a contribution from the repeatability of integrating the atomic signal:

$$\sigma_Y^2 = \sigma_{repeat}^2 + Y^2[(\frac{\sigma_{f_{dil}}}{f_{dil}})^2 + (\frac{\sigma_V}{V})^2] \quad (5A)$$

or, by using Eqn. 1 to substitute Y:

$$\sigma_Y^2 = \sigma_{repeat}^2 + (\frac{bC}{f} + Y_0)^2[(\frac{\sigma_{f_{dil}}}{f_{dil}})^2 + (\frac{\sigma_V}{V})^2] \quad (5B)$$

Values of the uncertainty components in Eqn. 2B and 5B were estimated by executing the corresponding step in the measurement procedure (Table 1). For example, the uncertainty of the volume of the test portion pipetted to the graphite furnace was estimated by pipetting 18 aliquots of aqueous solution containing radioactive ^{63}Ni. The variability in counts among the aliquots was taken as a measure of the uncertainty of the pipetted volume (after correction for the uncertainty of the counting process). Furthermore, the slope of the calibration curve (b) and blank signal (Y_0) was estimated as 0.0267 s·L·µmol^{-1} and 0.002 s, respectively. The correction factor, f, is presumably a function of the age of the graphite tube used to atomize the sample. An average correction factor of 1.015 was estimated from 168 results covering the concentration range 0.1-3 µmol/L.

Table 1 *Values of uncertainty components in determination of lead in blood by atomic absorption spectrometry. Atomic signals are obtained by peak integration and are expressed in seconds.*

Uncertainty component (standard deviation)	*Value*
Repeatability of sample signal, σ_{repeat}	0.00039 s
Uncertainty of volume pipetted to graphite furnace, σ_V/v	0.75%
Uncertainty of dilution factor, σ_{fdil}	0.09%
Uncertainty of slope, σ_b/b	1.8%
Uncertainty of correction factor, σ_f/f	0.86%
Uncertainty of blank signal, σ_{Y_0}	0.00054 s
Uncertainty of sample signal, σ_Y	0.00039 - 0.00052 s (*)

(*) The standard deviation of Y, σ_Y, is a function of lead concentration (Eqn. 5). The range in the Table covers 0.15-1.63 µmol/L

4 STATISTICAL CONTROL

An analytical method is in statistical control when its known sources of uncertainty fully account for the observed variability. This situation can be documented by carrying out an analysis of precision.[4]

Considering the sources of uncertainty in Table 1 the anticipated (*a priori*) standard deviation of a measurement result is calculated by inserting into Eqn. 5B and 2B. The *a priori* standard deviation is compared to the observed (*a posteriori*) standard deviation obtained from measurements of 5 control materials (Table 2) obtained during one year. Thus, the measurement results were obtained under conditions between repeatability and reproducibility conditions ('intermediary reproducibility conditions').

Do the uncertainty components in Table 1 fully account for the observed variability? To answer this question we have to test if the *a posteriori* and *a priori* standard deviations, as given in Table 2, differ significantly.

Table 2 *Comparison between* a priori *and* a posteriori *standard deviation at various concentrations of lead in blood at 'intermediary reproducibility conditions'. ν is the degrees of freedom for the* a posteriori *standard deviation.*

	Standard deviation				
Concentration (μmol/L)	*A priori* (μmol/L)	*A posteriori* (μmol/L)	ν	T_j	Test
0.13	0.026	0.022	7	5.01	$P(\chi^2_\nu \geq T_j) > 0.40$
0.34	0.026	0.026	12	12.00	$P(\chi^2_\nu \geq T_j) > 0.40$
0.63	0.029	0.035	12	17.48	$P(\chi^2_\nu \geq T_j) > 0.10$
0.95	0.032	0.036	9	11.39	$P(\chi^2_\nu \geq T_j) > 0.20$
1.63	0.043	0.042	9	8.59	$P(\chi^2_\nu \geq T_j) > 0.40$
T=44.66, ν=49					$0.40 < P(\chi^2_{49}=44.66) < 0.60$

To test the standard deviations the test statistic T is calculated:[4-6]

$$T_j = \frac{\sum_{i=1}^{n} (x_{ij} - \overline{x_j})^2}{\sigma_j^2} = \frac{\nu_j s_j^2(\nu_j)}{\sigma_j^2} \qquad (6)$$

where s_j is the *a posteriori* standard deviation with ν_j degrees of freedom, and σ_j is the *a priori* standard deviation. The distribution of T_j approximates a χ^2-distribution with ν_j degrees of freedom.[7] Applying the addition theorem for the χ^2-distribution the following equation is obtained:

$$T = \sum_{j=1}^{k} T_j(\nu_j) \qquad \text{and} \qquad \nu = \sum_{j=1}^{k} \nu_k \tag{7}$$

which can be used to test overall deviation between *a priori* and *a posteriori* standard deviations. Like the individual T_j values, T approximates a χ^2-distribution ($\nu = \nu_1 + .. \nu_j + .. \nu_k$ degrees of freedom).

From Table 2 it is seen that neither T=44.66 nor the values of T_j differs significantly from their expected values, and it is therefore concluded the *a posteriori* standard deviations do not differ significantly from their *a priori* counterparts. Thus, the uncertainty components in Table 1 fully account for the variability of the measurement results.

5 IMPROVING METHOD PERFORMANCE

Estimation of measurement uncertainty from uncertainty components makes it possible to identify the most significant uncertainty components. The components are combined as variances (*i.e.* squared standard deviations), and usually the significant uncertainty components are small in number. In order to identify the significant uncertainty components it is noticed, that each term in Eqn. 2A expresses the partial contribution to the measurement uncertainty from one uncertainty component. For example, the relative partial contribution (p) from the uncertainty of the correction factor f is given by:

$$p_f = \frac{\left(\frac{\partial C}{\partial f}\right)^2 \sigma_f^2}{\sigma_C^2} \tag{8}$$

The relative partial contributions to the measurement uncertainty (variance) from uncertainty of the sample signal (Y), blank signal (Y_0) and calibration slope (b) are calculated by similar equations. The results are illustrated in Figure 2.

Figure 2 shows that the blank signal is the largest contributor to uncertainty at low blood lead concentrations. The uncertainty of blank signal measurement is due to integrating an absorbance signal that fluctuates randomly around the baseline; therefore, the contribution from this uncertainty component can be reduced by reducing the integration time (if possible), or by increasing the number of blank measurements. Reducing the magnitude of other uncertainty components will only lead to a minor reduction of the measurement uncertainty.

6 TRACEABILITY

Atomic absorption spectrometry, like most methods in analytical chemistry, relies on calibration with standards. Ideally, these standards are calibrated against reference standards, which provides the traceability to a primary standard with the highest metrological qualities, *e.g.* a pure element. Thus, the calibration procedure embodies the realization of the appropriate unit of measurement, *e.g.* the mole or the kilogram.

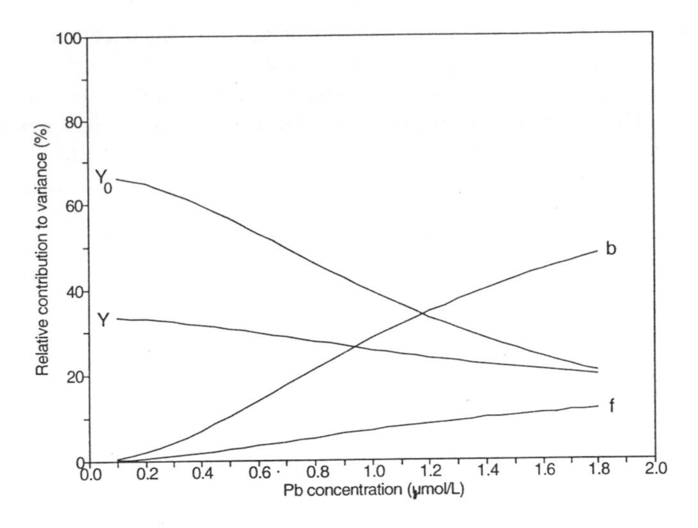

Figure 2 *Relative contributions from the uncertainty of atomic absorption signal (Y), blank signal (Y_0), slope of calibration curve (b) and correction factor (f) to the uncertainty of the concentration*

In the above analysis of precision only those uncertainty components that contribute to variability of measurements under experimental conditions were considered. Uncertainty of the realization of the basic SI unit (the mole) or the derived substance concentration unit (µmol/L) was not included, since this uncertainty component did not contribute to variability in this specific case. However, when reporting measurement results, *e.g.* C=0.26 µmol/L, reference to SI units are obviously made, and therefore the uncertainty from realizing the SI units must be included in the uncertainty reported with the result.

This uncertainty component may at first appear to be a metrological speciality with little relevance to analytical chemists. But in some instances it may actually be a significant source of uncertainty, *e.g.* in determination of complex biological macromolecules. Analytical methods for determination of hormones, proteins etc. in biological tissues (*e.g.* immunoassays) may be very precise, yielding excellent values of repeatability and reproducibility standard deviations, but the realization of appropriate SI units through calibration is very difficult.[8] In this case neither repeatability nor reproducibility standard deviations may provide a satisfactory uncertainty estimate suited for comparison of measurement results or decision-making.

Returning to atomic absorption spectrometry we therefore recognize that the list of uncertainty components in Table 1 is not complete. For example, with regard to the calibration curve only the uncertainty of repeatability is considered. Two additional groups of uncertainty components should be considered:[9] Firstly, the uncertainty of assigning values to the standards, and secondly, the uncertainty due to matrix effects. In this specific example the standards are produced by spiking a pool of whole blood with an aqueous lead solution. Therefore, the first group comprises the uncertainty on the certified value of the aqueous lead standard, uncertainty on the purity of the lead, uncertainty on balances and volumetric equipment used in the dilution procedure, and uncertainty arising from contamination of water used to dilute the aqueous standard. The second group comprises effects from the standards differing from real samples, *e.g.* regarding the chemical species and the matrix. The third group of uncertainties not included in Table 1 concerns effects directly related to the unknown sample, *e.g.* contamination or loss of element during collection or preparation of the sample, heterogeneity etc. The uncertainty from heterogeneity of the blood sample affects the uncertainty of the sample signal, σ_Y. Experience has shown that the standard deviation of the concentration may double in extreme cases, *e.g.* where a less effective anticoagulant has been used. For other analytical methods it is well-known that the strategy of sampling plays a significant role for the uncertainty.[10] Examples of uncertainty components contributing to measurement uncertainty, but usually not considered in experimental estimation of standard deviations, are given in Table 3.

Table 3 *Uncertainty components to be considered when estimating the uncertainty*

Uncertainty component	Example
Calibration	Value assigned to standard or certified reference material Realization of SI unit Purity of standard Balances and volumetric equipment Contamination/loss during preparation Matrix effects
Sample	Representativeness Contamination/loss at collection or preparation Homogeneity

7 IMPLEMENTING THE UNCERTAINTY CONCEPT

Reasons have been give above to illustrate the need for implementing and using the concept of uncertainty. The fundament is given by the *Guide for Expressing Uncertainty of Measurements* published jointly by ISO, IUPAC and other international organizations.[1] At least one interpretation of this guide addressed to the analytical chemistry community is under preparation.[11] In addition, national accreditation bodies at least formally require analytical laboratories to document the uncertainty of their measurement results. In reality, the accreditation bodies have to await the appearance of knowledge and expertise necessary for uncertainty estimation in the various accreditation areas.[12] This expertise is non-existing for the time being, at least in most areas. What can be done to change this state of affairs?

First, we suggest that the responsible international organizations, primarily IUPAC, take the consequence of being co-publishers of the above-mentioned guide. Currently, the policy of IUPAC concerning uncertainty is unclear.

It is suggested that IUPAC should:

- Inform the members of the organizations views and policy concerning uncertainty.
- Take the necessary action to carry out the above-mentioned policy.

In step with the society's need for comparable and reliable measurement data the need for a complete 'measurement' policy from IUPAC grows. This policy should address not only uncertainty, but all metrological aspects of analytical chemistry, *e.g.* traceability, the role of certified reference materials etc.

The actions taken by IUPAC with respect to an 'uncertainty policy' will of course depend on the actual policy. However, several important issues could be addressed:

- Development and dissemination of interpretations of the guide[1] for specific areas within analytical chemistry.
- Produce examples of uncertainty estimation.
- Support efforts to solve problems that are specific to the chemistry community, *e.g.* uncertainty from matrix effects, uncertainty of sampling etc.
- Initiate meetings and workshops focusing on uncertainty.

Other important aspects for implementing the principles of the guide[1] are management and education of laboratory staff. However, this is only possible if a clear commitment is demonstrated by the measurement community. Therefore the statement of an IUPAC policy concerning uncertainty is urgently needed.

References

1 'Guide to Expression of Measurement Uncertainty'. Publ. BIPM, IEC, IFCC, ISO, IUPAC, IUPAP, OIML, International Organization for Standardization, Geneva, Switzerland, 1993.

2 P. Giacomo. *Metrologia*, 1981, **17**, 73.

3 ISO 3534, 'Statistics - Vocabulary and symbols - Part 1: Probability and general statistical terms', International Organization for Standardization, Geneva, Switzerland, 1993.

4 K. Heydorn. *Mikrochim. Acta. Wien* 1991, **III**, 1.

5 K. Heydorn. *Anal. Chim. Acta* 1993, **283**, 494.

6 L. Lyons. *J. Phys. A: Math. Gen.* 1992, **25**, 1967.

7 'Geigy Scientific Tables', ed. C. Lentner. Ciba-Geigy, Basle, Switzerland, 1982, p. 202.

8 J.T. Whicher. *Scand. J. Clin. Lab. Invest.* 1991, **51 (suppl. 205)**, 21.

9 A. Marschal, 'Traceability and calibration in analytical chemistry and materials testing. Principles and applications to real life, in connection with ISO 9000, EN 45000 and ISO guide 25', *ISO/REMCO N 319*, International Organization for Standardization, Geneva, Switzerland, 1994.

10 J.M. Christensen, E. Olsen. *Fresenius. J. Anal. Chem.* 1991, **341**, 573.

11 'Quantifying uncertainty in analytical measurement', Eurachem Workshop draft (version 5), September 1994.

12 'The expression of uncertainties in testing (draft)', WELAC WG5, 1994.

Customer Requirements – The Essential Consideration of 'We' when Supplying a New Instrument System

J. P. Hammond and M. J. Long

UV MARKETING, ATI UNICAM, YORK STREET, CAMBRIDGE CB1 2PX, UK

1 ABSTRACT

The modern definition of an instrument no longer just refers to the spectrometer on the bench; the requirement to show and prove 'Quality of Installation and Design' are just two of the additional considerations. These new boundary definitions are discussed at length, with particular attention being paid to the implementation of current international regulatory standards by a 'Supplier Guidance' document. The paper follows a logical path from system design, through manufacturing, to installation, onward to training, and finally maintenance. This path is reproduced in the regulatory environments as the Qualification route through Design and Specification (DQ and SQ), Installation (IQ), Operation (OQ), and Performance (PQ), and it is the tie between these two parallel processes of instrument design and manufacture, and Regulatory Qualification that forms the major part of this paper and justifies the "We" inclusion in the title.

2 INTRODUCTION

Regulatory compliance is the major issue of this decade, currently being addressed by the pharmaceutical industry. This position as the 'Number 1' issue can be justified by the following three statements of fact.

Firstly, many major companies now have a Validation Manager, whose sole responsibility is to ensure compliance to the desired regulatory standard or standards.

Secondly, in February 1994, the UK Pharmaceutical Industry Computer Systems Validation Forum (PICSVF) issued the first draft of a supplier guidance document entitled

"Validation of Automated Systems in Pharmaceutical Manufacture" [1]

which fully describes the process by which a system is specified, designed, installed, and maintained; and describes the supplier / customer relationship.

This document sold over 600 copies before the release of the first version in March 1995.

Lastly, compliance failure can lead to delayed product introductions, the cost of which has been estimated at $100,000 per day, [2] restricted factory output and adverse publicity that harms the company image and depresses its share price.

The cost of obtaining and maintaining compliance is rising all the time as legislation is introduced in the push for higher standards. For example, to trade effectively on a global scale and sell into the USA, European pharmaceutical companies must comply to the requirements of the US Food and Drug Administration (FDA). The FDA continues to increase the number of foreign inspections it carries out per annum, increasing from 228 in 1993 to an estimated figure of 1100 in 1994, the number of drug manufacturers reviewed; [3] and in the process taking a much tougher line with producers.

With these requirements in mind, decisions can be made at the specification stage of a new product to ensure that it will meet the current mandates of the industry, accommodate future needs and thereby assist companies in their quest for regulatory compliance. Also, whilst the onus is on the drug manufacturer to prove the initiation and maintenance of control of a system throughout its life-cycle, by the provision of information and tools suppliers can assist in this process.

This paper will discuss some of those design considerations, and "life-cycle assistance"; but before the impact of the regulatory environment on instrument design can be discussed fully, we must clarify exactly what is meant by the 'Regulatory Environment'.

3 DESIGNING FOR THE REGULATORY ENVIRONMENT

The Regulatory Environment has a fundamental definition of a controlled working environment, audited by a third party, which then branches into several lower levels. This structure is shown in Figure 1.

At this top level, we find the recently introduced ISO Guidelines for Internal Quality Control.[4] These guidelines, constructed by a joint IUPAC/ISO/AOAC Working Party, aim to harmonise and remove the obvious branching that currently occurs at the next level down.

However, it is from this top level that the auditing bodies of the regulators, the Regulatory Enforcement Agency e.g. Food and Drug Administration (FDA) in the USA, Medicines Control Agency in the UK, etc. work , so they also require evidence of compliance in all sub levels.

At the next level down, currently there is a basic split into two key types: The first branch is 'Quality and Environment', a major consideration, defined as having the structure in place to perform the analysis correctly; and the second major concept is 'Application and Accreditation', which refers to the act of performing the analysis confidently and correctly.

3.1 Quality and Environment

This requirement is usually fulfilled by recognised Quality systems, and the definition of Good Laboratory Practice (GLP) shows how it achieves this.[5]

" Good Laboratory Practice (GLP) is concerned with the organisational processes and the conditions under which the laboratory studies are planned, performed, monitored, recorded and reported. Adherence by laboratories to the Principles of Good Laboratory Practice ensures the proper planning of studies and the provision of adequate means to carry them out. It facilitates the proper conduct of studies, promotes their full and accurate reporting, and provides a means whereby the integrity of the studies can be verified. The application of GLP to studies assures the quality and integrity of the data generated and allows its use by Government Regulatory Authorities in hazard and risk assessment of chemicals."

In implementation, there must be produced clear and concise Standard Operating Procedures (SOP's). Instrument manufacturers can assist in this process by producing clear and correct operating instructions for systems, which can then be incorporated into the Users SOP.

The bottom line is that GLP exists to protect raw data, which is generally accepted as the first recorded representation of any item of data that is intended to support a regulatory submission , whether paper or electronic. This data protection is achieved by assuring that raw data remains unaltered, and records are kept of any data manipulations, e.g. smoothing a spectrum, in the form of an audit trail.

Clearly, from this requirement to preserve data integrity, comes a list of additional essential items: equipment identities, i.e. make, model and serial number, time/date of data generation, manipulation history via an audit trail, and permanent hard copy output. These requirements can easily be incorporated into instrument control software, only if the requirements are considered at the design stage.

ISO 9000 is the international standard for a quality management system. Laboratories in general will not seek ISO 9001 certification, but increasingly pharmaceutical and other manufacturing companies are being recommended to seek verification of vendors adherence to ISO 9000 whether it be raw materials supplies or system software. ISO 9000 registration does not guarantee quality per se, but it does place the requirement on the vendor to put in place useful documentation systems.

Referring back to Figure 1, if we consider the next level down, we find the topic of System Validation. This key area, and the symbiotic supplier / customer relationship required therein forms the main body of this paper.

3.2 Application and Accreditation

Accreditation is the procedure by which a laboratory is assessed to perform a specific range of tests or measurements. The accreditation covers the range of materials tested or analysed, the tests carried out, the method and equipment used and the accuracy or precision expected, and is specific to the facility and the test.

Accreditation in Europe is based on guidelines set out in the European Standard "General Criteria or the Operation of Testing Laboratories" EN45001 and ISO IEC Guide 25 "General Requirements for the Competence of Calibration and Testing Laboratories". The guidance is applicable to the performance of all objective measurements, whether routine, ad-hoc or as part of research.

Most national laboratory accreditation schemes have based their standards on the international document ISO / IEC Guide 25. In Europe this is provided as EN 45001, and in the UK as BS7501. This standard is implemented in the UK by The National Measurement Accreditation Service (NAMAS), which was formally established by the UK Government in 1985. In June 1994, NAMAS and its partners in the Western European Laboratory Accreditation Co-operation (WELAC), combined with the Western European Calibration Co-operation (WECC) to form the European Co-operation for Accreditation of Laboratories (EAL). This harmonisation to date has resulted in the following list of countries participating in bilateral and/or multilateral agreements with NAMAS:

Australia, Denmark, Finland, France, Germany, Hong Kong, Ireland, Italy, Netherlands, New Zealand, Norway, Spain, South Africa, Sweden and Switzerland.

"How can instrument manufacturers assist with EAL compliance?"

The following five suggestions were given as the considered response from a co-ordinator and an assessor within the NAMAS organisation.

* User-friendly work instructions.
* Easy self maintenance.
* Advice on performance tests - what should you do to check?
* Filters traceable to International Standards (not own company standard).
* Ease of calibration.

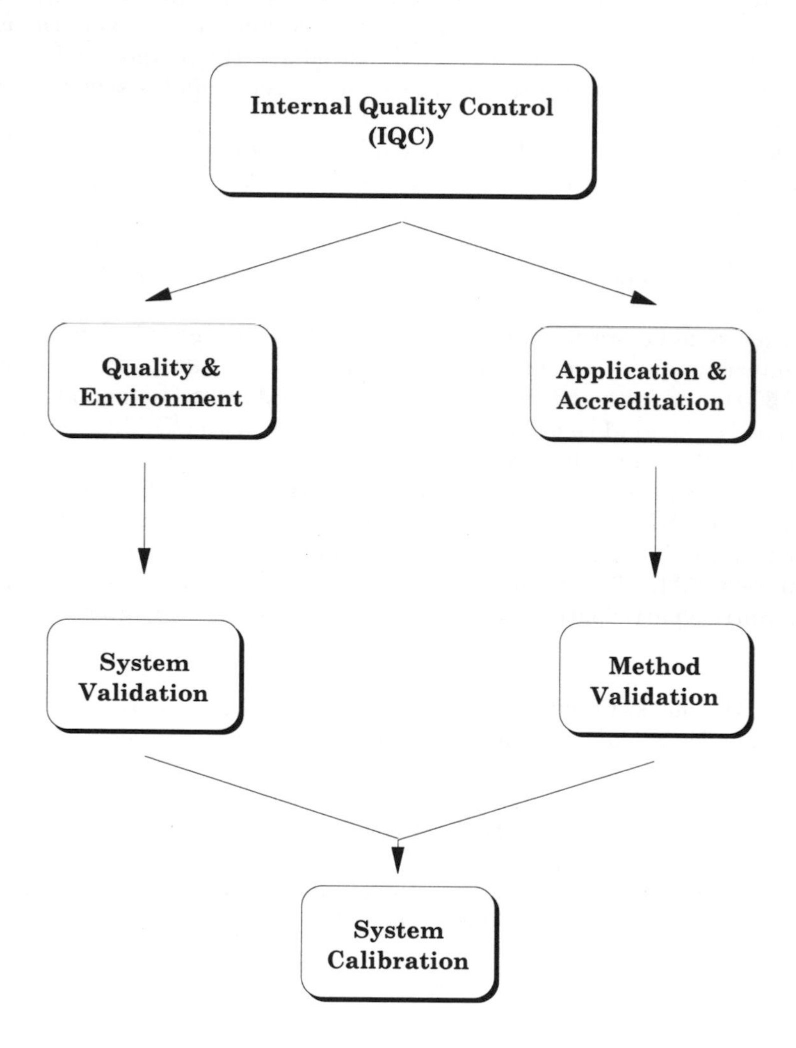

Figure 1 *The Regulatory Environment*

3.3 Method Validation

Method validation ensures the integrity and quality of the analytical method and has its own set of requirements. Manufacturers can help with this process by being able to provide full structural verification for software and that the application calculations have been tested and verified.
A typical set of analytical parameters required by the regulatory bodies is shown below:

* Precision
* Accuracy
* Limit of Detection
* Limit of Quantitation
* Selectivity (Specificity)
* Linearity and Range
* Ruggedness

Ruggedness requires documented evidence that the method under consideration is capable of being performed at different times, on different instruments etc.; the major point being the use of the word 'different' in each phrase. One would pose the question, would you even start to consider the results from two independent systems unless both could be shown to be suitably calibrated?

In fact, one could broaden the argument further by stating that all these parameters, can only be satisfactorily controlled if the instrument(s) are proven to be within calibration.

Having defined the regulatory environment in which we are working, we now need to clarify the terms 'instrument', 'design' and 'system'.

4 DEFINITIONS

4.1 Instrument or system?

The modern definition of an instrument no longer just refers to the spectrometer; the days of a self contained unit sitting on a laboratory bench with only a power cord connection, where the operator writes down the results by hand, has disappeared into the mists of time.

Until the age of the Personal Computer, a UV-Visible spectrometer fulfilled its dictionary definition as a "measuring device" in the form of a self contained box - a combination of optical and mechanical components, controlled by electronics. However, with the advent of the PC the boundaries have become enlarged, and less well defined and UV-Visible System is now probably a better terminology to use in place of instrument.

4.2 What do we mean by design?

Design in the context of this paper not only describes the process of producing an instrument on the bench, but also designing the surrounding support services to effectively fulfil its position in the regulatory space described above.

5 QUALIFICATION IN THE REGULATORY ENVIRONMENT

Figure 2. shows the generic path of the validation process within a typical regulatory environment.

5.1 Specification and Design Qualification (DQ & SQ)

Clearly, the development of a system is the first task that has to be "done properly"; a logical necessity; the Qualification requirement is reinforced by regulators in both the USA and UK requiring evidence of

* the use of rigorous design and specification methods
* full documentation, through Quality Control and Quality Assurance procedures.
* the use, at all times, of "suitably qualified and experienced" personnel.
* comprehensive, planned testing of all levels of the system.
* the application of stringent change control, error reporting and corrective procedures.

These arguments are based on the premise that you cannot test quality into a system, any more than you can test efficacy into a drug. It must be designed and built in from the outset.[6]

It is worth noting that these requirements apply, regardless of the basis on which the project is being conducted, whether instrument hardware or computer software. Whilst these requirements can be established from first principles, registration in a suitable approved quality scheme, e.g. ISO 9001 will provide the required structure, as will be shown later.

5.2 Installation Qualification (IQ)

Installation Qualification involves the checking of equipment and control system against the supplier standards of operating environment, physical connection, safety parameters and functional parameters prior to the initial utilisation of the system. In other words confirming that the system is properly installed.

5.3 Operational Qualification (OQ)

Operational Qualification is the process of demonstrating that the equipment will perform consistently as specified over all intended ranges. At this point, if we look back at Figure 2. the protocol takes inputs from the Environment/Quality route, in the form of Calibration, and Standard Operating Procedures.

5.4 Performance Qualification (PQ)

Performance Qualification is the testing of normal operation when manufacturing product after the IQ/OQ work is complete.

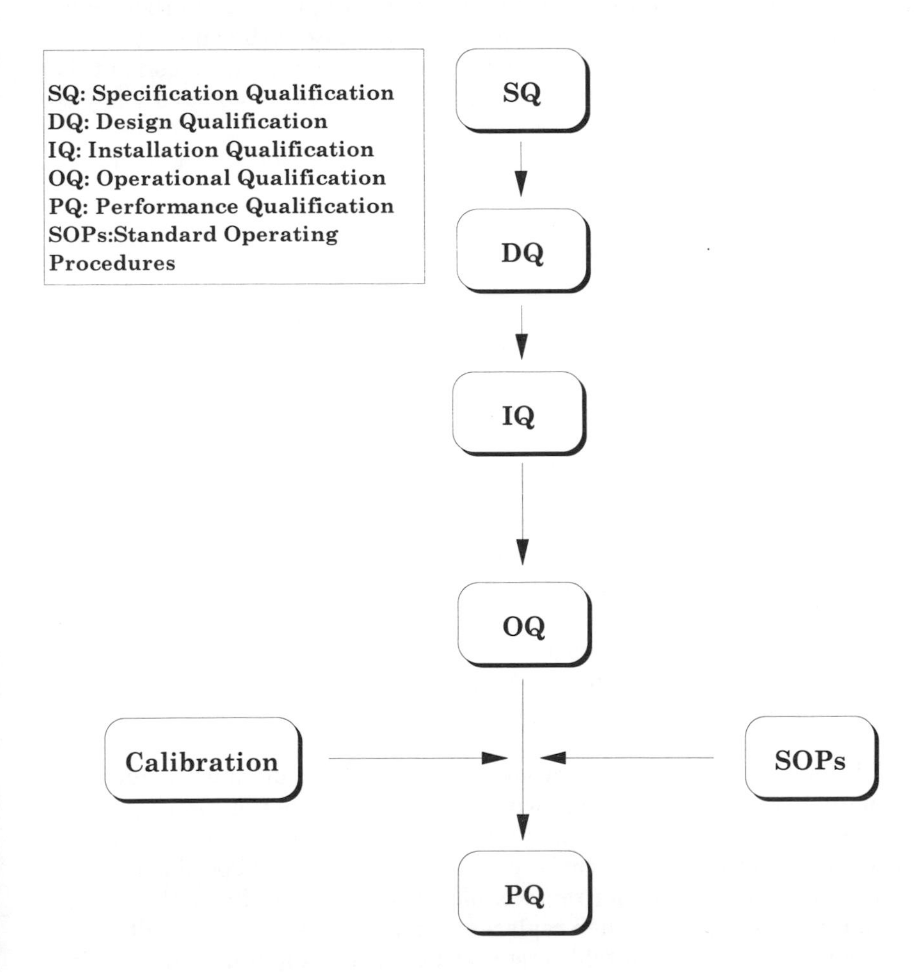

Figure 2 *A typical regulatory environment*

6 SYSTEM VALIDATION

As we have now defined the generic Qualification processes, Figure 3 shows these superimposed on a computer based instrument system. From the pathways shown, it can be seen that the validation of automated equipment, i.e. equipment fitted with a computer or PC follows a life cycle between that of a computer system and the spectrometer.

It can be conveniently sub-divided into the three key areas of 'Instrument Design & Manufacture', 'Computer Software & Control', and 'Qualification at the Customer site'. However, the division between the first two parallel processes and the on-site Qualification should not be viewed as in Figure 3, i.e. a dividing line mid-way between the supplier and customer. Historically, this may have been the philosophy adopted by the supplier, but in the current environment a more realistic division of responsibilities is shown in the modules that constitute these areas, and in these we see the diffusion of this boundary in both directions; only at the extremes are these modules solely the responsibility of either the supplier or customer.

In the following paragraphs, the 'Supplier' input into each of these domains is discussed.

6.1 Instrument Design & Manufacture

At this stage, two key requirements must be taken into consideration. Firstly, as has already been stated, quality must be designed in, it cannot be added at a later stage; and from this quality of build comes reliability, an essential requisite in any working environment. Quality comes not only from working within a quality system, e.g. ISO 9001, but also by the use of following 'components' when considering a UV-Visible spectrometer.

* Quartz coated optics
* Master blazed interferometric grating
* Precision quality mechanics e.g. wavelength drive, etc.
* Self aligning optics
* Rigidity of the optical bench
* Experience - commercial UV-Visible spectrometers have been available since 1949.

In addition, the design considerations for manufacturing must include the aim to manufacture a spectrometer with as small a number of component parts as practicable; the smaller the number, the less chance there is for things to go wrong; and as a secondary consideration, where possible these parts are combined as customer replaceable modules, to minimise the 'down-time' of an installed system when replacement is required. Once built and tested, the production of a suitable test certificate showing compliance to the specification will assist the customer in proving the quality of the installation.

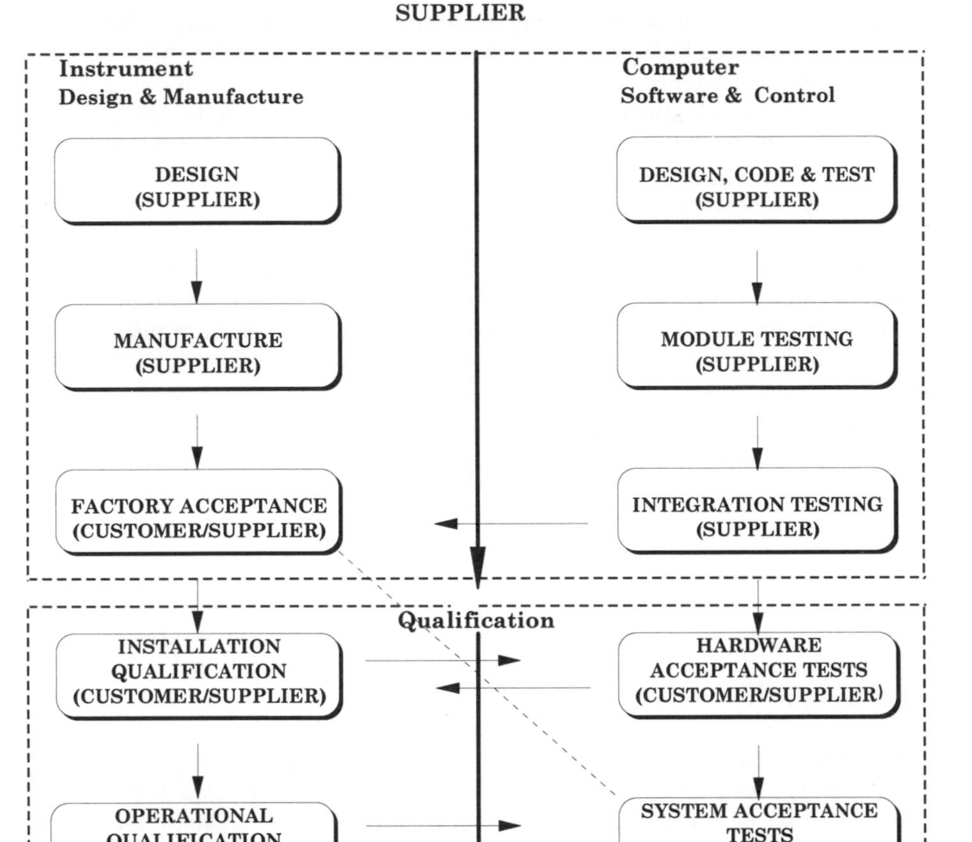

Figure 3 *Instrument System Validation*

Our spectrometer now has a certificate of 'Compliance to Specification' from the production area, and we can assist a regulatory customer further by effectively commissioning the complete system. Evidence of this system testing can be provided in the form of a Commissioning Certificate. Being the final supplier process before the customer based Installation Qualification, it provides added confidence for a trouble free installation, and as a secondary benefit, use of the Installation Manual and Check List provided to the Customer shows up any errors or omissions in these supplied documents.

6.2 Computer Software & Control

In an ISO 9001 environment, this is effectively covered by an overview document that specifies the 'Software Life-cycle' and all the associated documented Local Procedures required to implement it. For example, modules are tested to a defined script by an independent party as defined by a Local Procedure. Once this level of testing has been achieved, a formal Software Acceptance Test, again to a defined script is undertaken by a person external to the group.

In addition to the actual testing of the software there are procedures relevant to problem reporting, change control, archiving, etc.

6.3 Qualification

In this area the essential requirement is to provide documented evidence of the correct installation (IQ), initial operation to specification (OQ), and performance in "normal use" (PQ).

By following the guidelines as specified by the PICSVF,[1] suitable plans for each of these areas can be written, and provided in the form a Log Book. If these plans are written by the supplier, they ensure the correct procedures are followed, whilst removing the burden of fabrication from the Customer. A completed Log Book not only provides evidence of the Qualification process, but acts as a central repository for all the additional information gathered during the life-cycle of the system.

6.3.1 Installation Qualification. An appropriate plan will document correct observation of safety requirements, unpacking, and connection together of all the component parts of a system.

6.3.2 Operational and Performance Qualification. If the regulatory environment is considered at the design stage, then any accessories that assist the validation process can be easily incorporated. As shown in Figure 1, the common denominator in any validation process is the fundamental requirement to calibrate the spectrometer. One way of achieving this is to automate the process of placing calibration filters in the sample beam by mounting these on a motor driven wheel. Figure 4 shows a typical unit of this type.

When mounted inside the dust free optical compartment of an instrument, the chances of the filters being affected by external considerations are much reduced.

As a reputed spectroscopist once said;[7]

" ... The consideration of tolerances on absorbance checking filters pales into insignificance if one considers the effect of a thumb print on the measuring face."

Returning to the design of the Self Test accessory shown in Figure 4, the filters should be selected to test the specification range of the instrument. If these are then calibrated traceable to international standards, Operational Qualification becomes the simple exercise of running and recording the results obtained.

Performance Qualification involves repeating this process, throughout the operating life of the spectrometer; by the provision of this unit, and suitable software to interpret the data produced, valuable trend analysis can be performed.

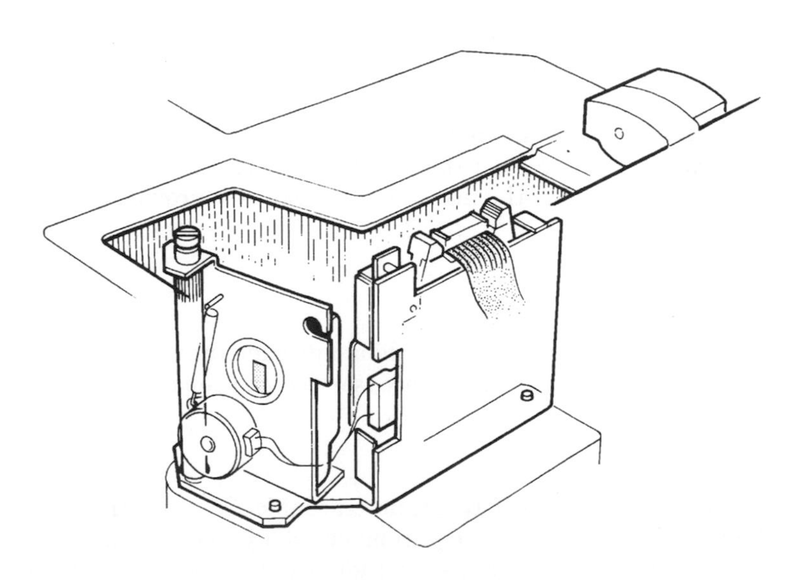

Figure 4 *A typical instrument Self Test unit.*

6.3.3 System Acceptance Tests. In their Supplier Guidance document already mentioned, the UK PICSVF states that System Acceptance Tests can be split into two, the first part of which can be achieved by the Factory Acceptance Tests, the second by defined on-site testing. Therefore, by providing these in a suitable format, the manufacturer can considerably reduce the amount of System Acceptance Testing required of the customer, by making a copy of the factory system tests available and by providing the blank Software Acceptance Test protocol. This assistance is shown in Figure 3 by the dotted line that traverses the supplier / customer boundary.

6.3.4 System Performance Qualification. A philosophical and practical point is, when does an analytical instrument become a computer system for validation purposes and vice versa. Factory Acceptance Tests are used to demonstrate that an instrument is capable of working within pre-defined limits. One approach is to extend this test to include the data system, so that for example a spectrum is obtained and the peak value calculated by the data system; the final result should be within acceptable limits. This is a "black box" approach to validation, but it provides a comparatively simple way of checking a system on a regular, daily basis, if required.

Another, more precise approach is to scan a reference sample, using on-board software (if available) and then reproduce the trace using the PC based system; clearly, within the limits of the system, the data obtained by both routes should be identical.

7 CONTINUING SYSTEM PERFORMANCE QUALIFICATION - MAINTENANCE OF THE LIFE-CYCLE

Now we have a fully validated new system, and it becomes the customer's responsibility to ensure that this validation is maintained throughout its working life. However, the manufacturer can still provide assistance in the following areas.

7.1 Documentation

Adequate, clear user manuals should provide fundamental information relating to calibration, maintenance, service and repair.

7.2 Calibration

As we have seen, a Self Test unit, traceable to international standards can fulfil most of the day to day requirements of a UV-Visible System. In the case of wavelength calibration however, there exists a more fundamental procedure - the use a low pressure mercury discharge lamp. It is used where high accuracy calibration is required,[7] and is quoted as "The best single source of ultraviolet and visible calibration spectra."[8]

Essentially, the procedure is to replace the Deuterium source in a UV-Visible spectrometer with a Mercury Pen, and then measure the emission spectrum produced. Figure 5 shows one solution to the normal problem of removal or refitting of the Deuterium lamp. With this configuration, the Mercury Pen can be quickly and easily located, selection being achieved under software control using the Lamp Change mirror. Again this desired system capability has only been fulfilled by consideration of the requirement at the initial design stage.

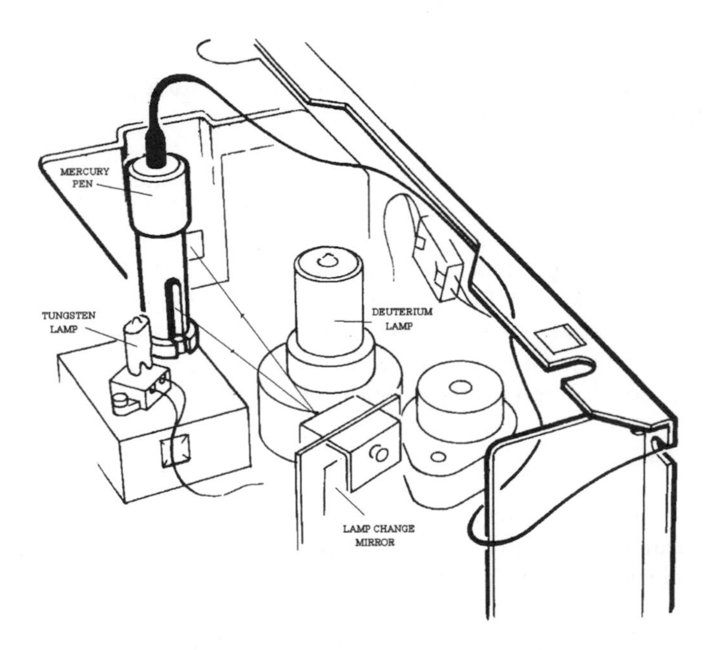

Figure 5 *A Mercury Pen lamp installed in a modern spectrometer*

7.3 Maintenance, Service and Repair

The User manual details when and how to service the spectrometer. As has already been stated, the requirement to provide documented evidence of maintenance, error logging, and change control can be effectively combined with the Qualification plans in a system Log Book. Further assistance can be provided by the supplier in the form of a spectrometer Service Manual, telephone support, trained Service Organisation and User Service training courses.

7.4 Training

In the 'Quality and Environment' area of regulatory compliance the use of adequately trained staff is specifically detailed, with reference to the inspection of training records, programs, etc.[5]

Whilst most customers will have their own individual requirements, etc. the supplier can assist the process by providing; suitable tutorials in the PC based software, suggested training plans, etc.

8 CONCLUSIONS

" The 1990's are the decade of quality. Laboratories must publish quality information or they will not exist" [9]

This one sentence summarises the environment in which we all now find ourselves and unless system design takes account of this new climate, it will not only be the laboratories that cease to exist.

However, as this paper has shown, by incorporating the regulatory requirements from the very first design concepts, a UV-Visible system that is "adequate for its intended use" can be produced; and perhaps more importantly can be shown by the use of its own Log book to produce data that is "fit for purpose" throughout its working life.

References

1. 'Pharmaceutical Industry Supplier Guidance: Validation in Pharmaceutical Manufacture', UK Pharmaceutical Industry Computer Systems Validation Forum, Version 1.0, Issue A, March 1994.

2. Dr. Sandy Wienberg, 'Case Studies in System Validation: Sixteen Theses', Weinberg Associates, Inc.

3. Wechsler, 'Fairness & Foreign Policy', *Pharmaceutical Technology Europe*, Vol. 6, No. 9, October 1994, p. 16.

4. IUPAC/ISO/AOAC Working Party, 'Harmonised Guidelines for Internal Quality Control in Analytical Chemistry Laboratories', *N271 Rev.*, Nov. 1994.

5. 'Good Laboratory Practice - The United Kingdom Compliance Programme', UK Department of Health Publication, 1989, p. 3.

6. Murphy, M, 'LIMS and the Regulated Laboratory', Conference Proceedings, PharmAnalysis Europe, Edinburgh. 11-12 Oct. 1993, Advanstar Communications, Chester, UK, p. 143.

7. Burgess, C. and Knowles, A. (eds.) (1981), 'Standards in Absorption Spectrometry', Chapman and Hall, London, chapter 7.2.

8. 'U.S.P. Pharmacopoeia XXI 1985', United States Pharmacopeial Convention, Inc, Rockville, p. 1273.

9. McDowall, R.D., 'Data Integrity from Sample Preparation to Report: Automation in a regulated environment', Conference Proceedings, PharmAnalysis Europe, Edinburgh. 11-12 Oct. 1993, Advanstar Communications, Chester, UK, p. 57.

COMAR – The International Database for Certified Reference Materials

Harry Klich

FEDERAL INSTITUTE FOR MATERIALS RESEARCH AND TESTING, RUDOWER CHAUSEE 5, 12489 BERLIN, GERMANY

1 SUMMARY

With more than 200 producers of reference materials (RM) throughout the world, it is often difficult to find the best reference material for a specific application. The database COMAR has been developed to assist scientists in finding the reference material they need.

2 INTRODUCTION

The need for accurate measurments is becoming ever more important as science and society become more complex and more demanding. In order to reliably achieve such measurements, it is recognised that an analytical laboratory should implement the requirements of a recognised quality assurance system, underpinned by third-party assessment, use validated methodology, participate in proficiency testing schemes where appropriate, use certified reference materials (CRMs) and employ adequately trained analysts.

Of these, the use of CRMs is probably the most important single requirement because CRMs act as the traceability link to the SI international system of measurement. By the application of a CRM whose matrix and analyte composition match as closely as possible that of the samples under test, it is possible for the analyst to assure himself that the measurements have been properly carried out to the requiered level of accuracy.

Meanwhile the database provides information about more than 9000 CRM's, coming from France, USA, Germany, United Kingdom, Japan, CIS, China and other countries. These data are organized in eight basic fields of application, each containing up to ten subcategories.

3 CONTENTS OF THE DATABASE

The structure of the database allows the user to decide first a main-category in order to narrow down the search. (*Figure 1*)

3.1 Fields of application

3.1.1 Ferrous reference materials

Pure metal RMs for the steel industry,
Unalloyed steels (Euronorm classification),
Low alloy steels (Euronorm classification),
High alloy steels (Euronorm classification),
Raw materials,
By-products,
Cast iron,
Special alloys (Euronorm classification),
Other metallurgical RMs for the steel industry.

3.1.2 Non-ferruos reference materials

Pure RMs for non-ferrous metallurgy
Lithium, beryllium, alkali and alkaline earth metals,
Aluminium, magnesium, silicon and alloys,
Copper, zinc, lead, tin, bismuth and alloys,
Titanium, vanadium and alloys,
Nickel, cobalt, chromium and refractory metals,
Precious metals and alloys,
Rare earths, thorium, uranium and transuranic elements,
Raw materials and by-products,
Other RMs for non-ferrous analyses.

3.1.3 Inorganic reference materials

Products of general interest and pure reagents,
Rocks, soils,
Glasses, refractories, ceramics, mineral fibres,
Building materials: cements, plasters,
Fertilizers,
Inorganic gases and gas mixtures,
Industrial acids and bases,
Oxides, salts,
Other inorganic RMs.

3.1.4 Reference materials for the quality of life

Environment,
Foodstuffs,
Consumer products
Agriculture: soils and plants,
Legal controls, criminology,
Other quality of life RMs.

3.1.5 Organic reference materials

Pure organic RMs,
Petroleum products and carbon derivatives,
Synthetic base products and large intermediates,
Common organics: solvents, gases and gas mixtures,
Plastics and rubbers, organic fibres,
Paints and varnishes, dyes,
Cosmetics, surfactants,
Pesticides and phytocides
Fine chemicals,
Other organic RMs.

3.1.6 Reference materials for physical and technical properties

RMs with optical properties,
RMs with mechanical properties,
RMs with electrical and magnetic properties
RMs for frequency,
RMs for radioactivity, isotopic radionuclides,
RMs for thermodynamics
RMs for physico-chemical properties,
Other RMs with physical and technological properties.

3.1.7 Biological and clinical reference materials

General medicine
Clinical chemistry,
Pathology and histology,
Haematology and cytology,
Immunohaematology, transfusion, transplant,
Immunology,
Parasitology,
Bacteriology and mycology,
Virology,
Other biological and clinical RMs.

3.1.8 Reference materials for industry

Raw materials and semi-finished products
Building, public works,
Transportation communications,
Electricity, electronics, computer industry,
Ores, mineral raw materials,
Measurement and testing techniques,
Fuels,
Other RMs for industry.

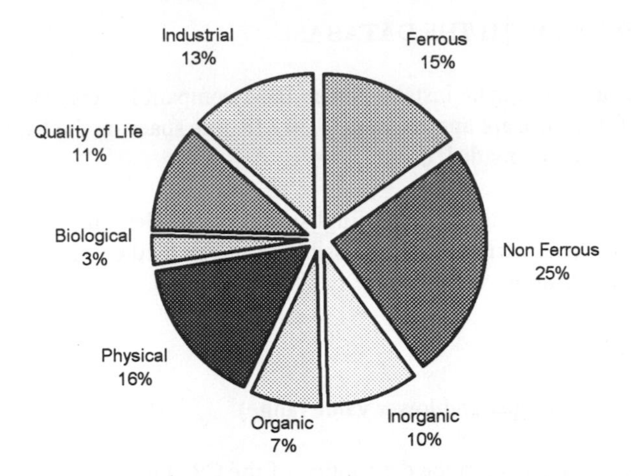

Figure 1 *Distribution by fields of application, based on 8850 CRMs (1994)*

3.2 For each CRM the database provides information on the following:

chemical composition (standard chemical notation)
molecular composition (CAS-RN)
physical properties (ISO Collection of Standards No.2)
conventional properties (standards of definition)
form, shape of samples. For example:

- *artifacts;*
- *cylinders and mushrooms;*
- *plates, discs;*
- *ingots, rough globules, bars and wires;*
- *crystals, monocrystals and bicrystals;*
- *chips, granules;*
- *powder;*
- *liquid;*
- *gas.*

country of origin;
producer (mailing address, telephone, fax, telex and contact name);
references (product code, comments).

4 AVAILABILITY OF THE COMAR DATABASE

Any chemical analysis and testing laboratory may wish to consider the use of the COMAR database in order to improve the reliability of its measurements or to assist in the development of new methodology. In order to achieve this objective, is is possible to contact the appropriate national coding centre, (Table 1) some of whom operate a free advisory service. However in order to obtain the full benefits available from the database, it is also available for purchase from the coding centres. In this way it is possible for the user to interrogate the database to a much greater degree than a single enquiry can achieve.

5 HOW TO WORK WITH THE DATABASE.

The COMAR database can be installed on an IBM-compatible personal computer. The requirements for the use are approxemately 20 MB free space on the hard disk, 520 kB RAM and a 1.4 MB floppy disk drive.

The database can be operated in either English, French or German. Retrieving information from the database is userfriendly. It is possible to search by the following criterias:

- fields of application
- form of the CRM
- country of origin
- producer
- certified value (upper and lower value range)

Furthermore a string search in the description of the CRM is available, so it is possible to use key words like SOIL, SEDIMENT, FLOUR.
In all cases the user can operate the database with AND/OR/NOT modes. A online help for the selection of molecules, physical and conventional properties, producers and countries during the interrogation is available.

It is also possible to extract data for special problems. As an example we extracted from the original data records all CRMs from the field "Agriculture, soils and plants" where Cd in soil has been certified. (Figure 2) This sample shows that a sufficent number of CRMs in a wide range is available. The uncertainty of the values is given as an errorbar. So it will be possible to estimate the probable uncertainties of new CRMs in a similar range. Also gaps, which represent the need for new reference materials, can be observed.
The dashed lines in the diagram represent e.g. the legal limits or guidelines of a German guide. (Source of limits: Berliner Liste, limits for inorganic and organic chemicals and harmful pollutants.)

6 CONCLUSION

The database COMAR is a useful tool to obtain different statistical information from all fields of certified reference materials. With simple programs it is possible to extract and visualize information for specific purposes, e.g. the availibility of CRMs in the field heavy metals in foodstuffs or soil. The need to develop new reference materials in several fields can be observed. For the production of new CRMs it will be possible to estimate the probable uncertainty in a similar range. Specific listings, sorted by fields of application, producers or country of origin can be created easily. The actuality of COMAR is assured by the co-operation of several institutions, which are responsible for CRMs in their country.

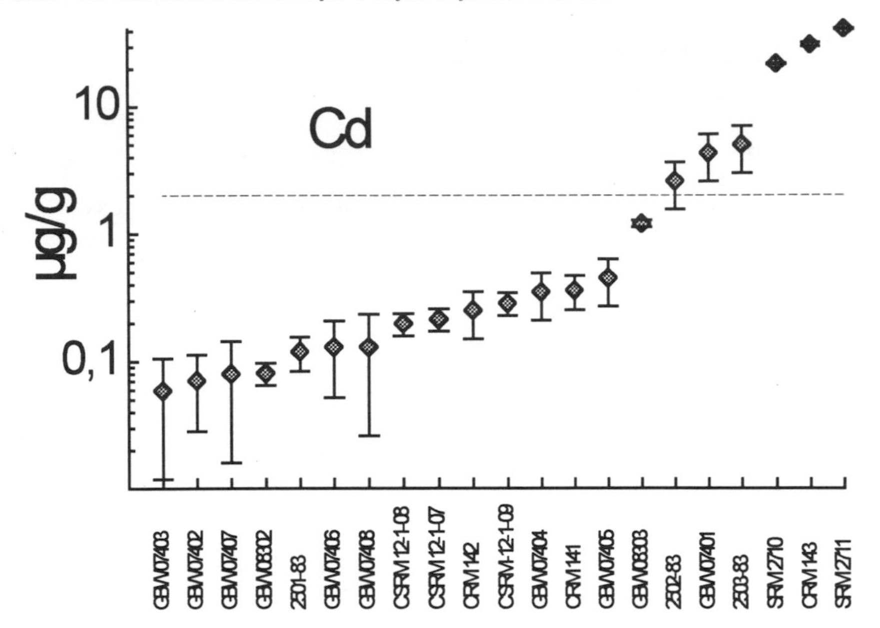

Figure 2 *Cd in soil*

Table 1 *Points of contact.*

Germany BAM
Bundesanstalt für Materialforschung und -prüfung Referat 10.01
Rudower Chaussee 5
D-12489Berlin

United Kingdom LGC
Laboratory of the Goverment Chemist
Office of Reference Materials
Queen´s Road Teddington
Middlesex TW11 0LY

Peoples Republic of China NRC-CRM
National Research Centre for CRM
No.7 District 11,
Hepingjie, Chaoyangqu
100013 Beijing

France LNE
Laboratoire National d'Essais
1, Rue Gaston Boissier
75015 Paris

USA NIST
National Inst. of Standards and Technology
Office of Standard Reference Materials
Gaithersburg, MD 20899

Japan IIII
Intern. Trade and Industry Inspection Institute
49-10,2 Nishihara Shibuya-ku
Tokyo 151

RUSSIA UNIIM
Ural Research Institute for Metrology
Krasnoarmeiskaya Street,4
620219 Ekaterinburg ,GSP-824

References

Klich H, Walker R (1993) Fresenius Z Anal Chem 345: 104-106

TQM Comes to the Laboratory

M. J. Allison

STOCKDALES PTY. LTD., CONSULTANTS TO INDUSTRY, BRIGHTON, VICTORIA 3086, AUSTRALIA

1 WHAT IS TQM?

What is TQM and how does it apply to laboratories? If that question was asked before 1994, it would have been difficult to find an answer. There was no accepted definition of the term, although many "Gurus" would offer their own interpretation. ISO 8402:1994 offers the following definition:

> "management approach of an organisation, centred on quality, based on the participation of all its members and aiming at long-term success through customer satisfaction, and benefits to all members of the organisation and to society"

This definition has been well thought through. There are also several notes as part of the definition. Let's explore how this definition fits with the modern world.

TQM applies to all levels and all functions of an organisation. It is certainly not the exclusive domain of the quality department, although convincing some people of this can be difficult.

TQM applies to all things these various levels do. Traditionally, we may have thought of quality being something that belongs to manufactured hardware. While this was the first area of application, today, it applies to marketing, managerial functions, design, processes, as well as hardware and software. The role of management in TQM is singled out for particular attention. Leadership by top management and training of all levels to enable people to work in a quality way, are twin keys to long-term success. Leadership is emphasised in many facets of human relations, but industry seems to be making the running. Without leadership, the organisation is a rudderless ship. On the other side of the coin, employees who are not properly trained to carry out their functions cannot translate good leadership into practical success.

The concept of quality in the TQM context is that it applies to all managerial objectives. It's hard to think of an objective to which it does not make sense to apply quality. Quality of service to the customer; quality of training for our people; quality of planning for our processes; quality of design for products and services; quality of conformance for manufactures and services; quality of delivery; quality of after-market support.

Benefits to all members of the organisation and society bring TQM into the mainstream of modern thinking about quality. We are all consumers and most of us are employees of some sort. Two Japanese quality teachers have given us concepts which put benefits into perspective.

Professor Noriaki Kano developed the concept of multi-dimensional quality. He speaks of "expected" quality, which if absent, will cause serious dissatisfaction. For example, we expect clean cutlery in a restaurant. This expectation does not have to be stated by the customer. Another kind of quality is "delighting" quality, where the supplier provides something extra, such as iced water on the table. The customer may not specify such a service characteristic, but it delights the customer and places the provider in a superior position to competitors. Of course, today's delighting quality may be tomorrow's expected quality. That is certainly the case for iced water in the U.S.A., where a restaurant would soon be empty if it failed to provide that service.

Genichi Taguchi introduced the concept of quality as "loss to society". Any departure from the required quality is described as a loss. This has had quite a profound effect on how the West views quality. For example, simply meeting a specification range may involve considerable loss. Product at the desired specification performs in a superior manner to product which is close to either the high or low limit. Of course, the extent to which this is true and the validity of the underlying mathematical model are subjects for debate, but it does make sense. Here is an example. An American TV manufacturer was in the habit of adjusting all sets so that the colour was always within a specified range. Nothing was shipped outside that range. A Japanese manufacturer developed a similar set, but made no adjustments. Some of the sets fell naturally outside the limits set by the American manufacturer. The customers thought the Japanese TV was superior. Figure 1 shows why this is so.

Although some of the Japanese sets had more loss (departure from ideal) than the American sets, *most* Japanese sets had smaller loss.

2 THE ROLE OF QUALITY ASSURANCE

Often, there is confusion about the role of Quality Assurance as opposed to TQM. QA can be regarded as a sub-set of TQM. It is generally agreed that without QA, it will be difficult to hold the gains made through TQM. Figure 2 shows the way in which QA locks improvements through TQM into place.

So how does QA do this? Without exhaustively examining all 20 elements of ISO 9001, suffice it to say that control mechanisms exist to ensure that organisation, information and methods are working as intended. Three "self healing" mechanisms exist within ISO 9001:

- internal quality auditing
- corrective and preventive action
- management review

It may be said that calibration of instruments is another form of self healing mechanism. Some organisations represented here today will of course have been certified to ISO 9000 series and also hold NATA registration as a laboratory. In both cases, the disciplines of these registrations contribute strongly to making sure that systems of operation work with minimum error and that errors are contained and corrected.

It should be noted that ISO 9000 series quality systems are applicable to hardware, software, processed materials and services.

Colour Distribution in TV Sets

Japanese sets

U.S.A. sets

D C B A B C D

A, B and C are diminishing grades of acceptability. D is outside specification.

Figure 1 *Loss function for a colour TV set*

Quality Assurance & Total Quality Management

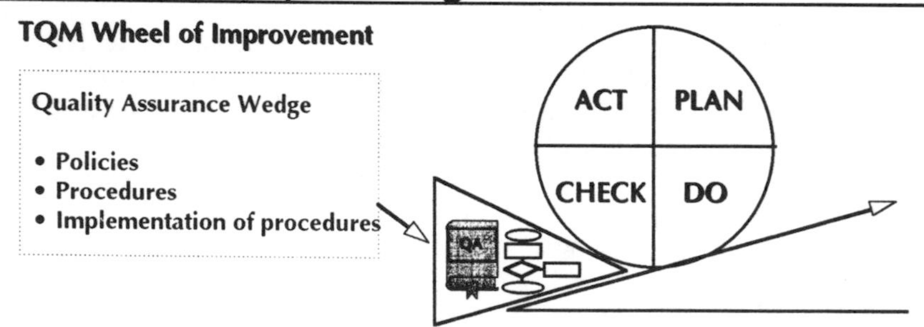

Continuous Improvement Imperatives

- **Understanding customer needs**
- **Improvement in systems**
- **Co-operation with suppliers**
- **Managing variation**
- **Involvement of employees**
- **Planning**

Figure 2 *The role of Quality Assurance in TQM*

3 CONTINUOUS IMPROVEMENT

Many people are under the impression that continuous improvement is what TQM is all about. Continuous improvement is an important element of TQM, but it is not the only element.

There are two main mechanisms by which continuous improvement takes place. One of these is through natural work groups, sometimes referred to as quality circles. These teams consist of people who naturally make up the group working closely in a defined area. For example, a quality circle might be formed in a laboratory, to look at ways of improving a test procedure. The group is familiar with the process and has a high level of skill in the subject, but not necessarily a high level of skill in problem solving and interpersonal behaviour. Such a group is generally restricted to addressing problems confined to the immediate work area. In Japan, extensive training of natural work groups has made them very effective in improving quality[1].

The other type of group is the cross-functional team. Such a team consists of people drawn from multiple departments and multiple levels of management. It represents the people involved in a typical process which cuts across many different parts of the organisation. This type of team can deal with much broader issues than the quality circle.

A more recent expansion of the cross-functional team is in the application of Business Process Re-engineering (BPR), which can be described as the "radical redesign of processes to achieve dramatic improvements in performance". Spectacular improvements have been made in many processes, because most processes have around 95% non value adding steps in them. BPR seeks to eradicate non value adding steps, through critically examining the intent of the process and redesigning the process as a "greenfields" exercise.

For Cross-functional teams, whether conventional or BPR, and to a lesser extent for natural work groups, the effectiveness of the improvement process depends on the chosen project being aligned with the business objectives.

4 TOOLS AND PLANNING IN TQM

When teams meet to make improvements, they usually apply a number of tools developed over the years for general problem solving. The use of these tools is also usually within a framework or methodology which calls for a sequential arrangement of activities which is expected to maximise the chances of success for the team.

Many of you will have heard of the "Seven Tools of TQC", which consist of:

Histogram	Scatter Diagram	Cause and Effect Diagram
Pareto Diagram	Check List	Run Chart
Flow Chart		

These old favourites have served us well, but increasingly, we need tools more oriented to prevention rather than solution of an existing problem. To this end, the "Seven Management Tools" enable both prevention and the handling of unmeasurable data. More about this later.

Of course, we shouldn't forget the importance of strategic planning in the TQM picture. This should be the basis on which the teamwork is directed. Otherwise, projects are arbitrary and without business direction.

In conventional strategic planning, we look at the competition, the markets (as they exist today), our strengths, weaknesses, etc. and formulate plans. All well and good - many organisations lack even this form of planning. But increasingly, we need to break away from conventional means.

The emerging approach is to *invent* new markets[2]. As we shall see, this is likely to be an important method for laboratories. The methods are based around competencies possessed and competencies needed to compete in the marketplaces of the future.

Statistical methods are important to discover the variation in processes and, in the case of more powerful methods recently developed[3], discover causes of problems. A good grounding in statistical methods is important in laboratories, because of the need to qualify analyses and to obtain best value in the case of exploratory work.

Believe it or not, laboratories are no different to any other service when it comes to looking at how TQM applies. Despite protestations to the contrary from the conservatives, laboratories do have customers. Most of you would have realised that decades ago. What's more, laboratories have both internal and external customers, just like other organisations.

Very few activities remain which still refuse to recognise the existence of customers. Even governments are beginning to realise that customers exist.

5 HOW CONCEPTS AND CHANGES AFFECT LABORATORIES

Let's consider the definition of ISO 8402 as it applies to a typical laboratory and see what is significant.

- centred on quality
- all members participate
- long-term success focus
- customer satisfaction
- benefits to members and society

All of these points seem quite reasonable and worthwhile. There are however, some changes in the wind and it is appropriate to dwell on these. They affect strategic planning and the type of problem solving needed to invent a successful future. The changes are all closely related to the bullet points above, often involving aspects of all five.

Here is a short table of changes which are happening in many industries and are relevant to laboratories (Table 1).

The human relations side of these predicted changes are considerable and will be addressed elsewhere in this symposium.

In the Seven Management Tools mentioned earlier, there is provision for making sure that things won't go wrong in the first place. The Process Decision Program Chart, mercifully abbreviated to PDPC, asks how facets of a process or product could fail and what countermeasures could be put in place to prevent that failure. This is not unlike the Failure Modes and Effects Analysis (FMEA) approach popular in the automotive industry and is used principally for design improvement, but may also be used for production.

Table 1 *Old and New Paradigms in Quality*

Old Paradigm	New Paradigm
"it will never happen again"	"it must not happen in the first place"
"the customers will beta test it for us"	"we must prove it out in the lab. - before the customer gets it"
"meet the spec."	"continuously reduce variation"
"it will be done on time"	"continuously reduce cycle time"
"please the customer"	"find out your customer's customer's needs"
"spend more time on the bench"	"spend more time with your customers"
"develop your analytical skills"	"develop your people skills"
"give an accurate description of the specimen"	"give an exploratory analysis for discussion"
"only use well established and reliable methods"	"experiment to find better ways"

A related technique, more concerned with repetitive activities, is Poka Yoke or mistake proofing. In this technique, simple devices are used to ensure that specified errors physically cannot happen. An example of this is the use of non-interchangeable fittings for oxygen and CO_2 lines in hospitals. Clearly, Poka Yoke is useful for ensuring that routine laboratory processes are error free.

Other management tools are used for planning and problem solving for the situations often faced by managers, where measurements are not available. Managers must have means to deal with anecdotal, chaotic and emotional information, despite the recent emphasis on "facts and data". The seven management tools provide such a capability.

Without doubt, laboratories will be called upon to play an increased role in minimising the risk associated with new products. Product liability cases and greater awareness of environmental issues will ensure this change takes place.

Already, reduced cycle time will be familiar to many, as computerised and automated laboratory methods have already replaced many of the older and slower techniques. The demand for quick response will accelerate as the consequences of delays become more serious. As with many other industries, laboratories may have to go to the customer, rather than customers go to the laboratory. If you think this is outrageous, consider the sophistication of many portable analysis devices and compare to twenty years ago. One sad aspect of the need for portability of analysis is of course the spread of AIDS.

Other changes likely to affect laboratories involve new approaches to discovering customer needs and ensuring customer satisfaction. The notion of customer's customer's needs is but one of these.

6 TRAPS AND PITFALLS

Just about any project of significance will founder if there is no support from executive management. Despite the obvious nature of this statement, lack of support from executive management for TQM initiatives, continues to be a major problem. The program is agreed to, assigned to the quality manager or coordinator and promptly forgotten. If TQM holds out such promise, how can executive management not want to follow progress keenly and be involved? Since a significant part of TQM is strategic planning, failure of executive management to be involved is equivalent to executive management refusing to be involved in the financial management of the enterprise.

Another common pitfall is to train people in techniques, send them back to the workplace and wait for something to happen. It doesn't. Improvement has to be managed. Few people can implement improvement projects without some coaching. Facilitators are needed to do this.

7 CONCLUSIONS

TQM is here to stay, be it by any other name. Laboratories have a central role in meeting public expectations of greater safety in products, improved response time and high reliability. The many disciplines and techniques of TQM will be additive to the ever-increasing complexity of modern analytical techniques.

References

1 Kaoru Ishikawa, "Guide to Quality Control", Asian Productivity Orgainsation, Tokyo, 1986

2. Gary Hamel & C.K. Prahalad, "Competing for the Future", Harvard Business School Press, 1994

3. Dr. Juergen Ude, "Manhattan Process Control - An Australian Development", *Quality Australia*, December 1992, pages 46-48

AUSTOX: An Exercise in Quality Control

Peter Bowron

TOXICOLOGY UNIT, ROYAL NORTH SHORE HOSPITAL, MACQUARIE HOSPITAL CAMPUS, BADAJOZ ROAD, NORTH RYDE, NSW, AUSTRALIA

1 INTRODUCTION

The field of drug toxicology is one which covers laboratories with a wide range of technical expertise, ranging from simple immunoassays to various forms of chromatography and mass spectrometry. Samples are drawn from many sources, such as patients on drug treatment programmes, hospital overdoses, employment testing, corrective services and the defence forces. It is a field where possible errors in analysis can lead to serious medical, social and legal consequences. While proficiency programmes existed for the related field of therapeutic monitoring, it has only been in the last 15 years that the issue of quality control in drug toxicology has been seriously addressed.

1.1 Australian Urine Drug Toxicology Programmes - A Brief History

Up until the 1980's no significant proficiency programmes existed in Australia for laboratories performing drug toxicology. At the start of the decade a number of organisations began independently to address the obvious needs for quality control in this area. In 1980, the Alfred Hospital in Victoria provided a programme aimed initially at serum paracetamol monitoring but expanded to cover specimens encountered in hospital situations. The programme was targeted predominantly at users of Toxilab in Victoria, although a number of interstate laboratories participated. Difficulties were often encountered in collecting sufficient sample for a viable control. The programme discontinued in 1984.

Also in 1980, Toxilab itself began providing proficiency samples, open to all laboratories but again targeted to laboratories using that technique. Although a useful tool for Toxilab users, the programme is not tailored to patterns of Australian drug usage.

In 1986 the Royal Australian College of Pathologists and the Australian Association of Clinical Biochemists launched a pilot programme, followed by a fully functional programme the following year. The programme traditionally functions as a service to clinical laboratories.

In 1981, a programme was started by the Oliver Latham Laboratory in New South Wales. As the State Health Department reference laboratory for the testing of drugs of abuse in urine, the laboratory had access to urine from many thousands of samples. This programme, originally called ODAP (**O**liver Latham Laboratory **D**rugs of **A**buse **P**rogramme), was renamed AUSTOX in 1992. Starting with 9 laboratories in New South Wales, Queensland and the Australian Capital Territory, it rapidly expanded to cover major public health and hospital laboratories in all mainland states of Australia, and more recently to cover both the public and private sectors and laboratories in the Asian region.

AUSTOX has from the beginning aimed to bring together all types of laboratories who are working in the field, ranging from small clinical laboratories using a single analytical technique, through hospital pathology services and up to state reference laboratories performing to forensic standards. 33 samples are sent out to each participating laboratory each year (3 per month for 11 months), with a wide variety of analytes covering both drugs of abuse and routine therapeutic compounds. Most recently, the programme has begun to address problems in the medicolegal field by encouraging laboratories performing this type of work to quantitate a number of abused drugs. Additionally, the programme provides peer group communication through a newsletter and an annual conference.

2 AUSTOX RATIONALE AND FUNCTION

2.1 Types Of Drug Toxicology Proficiency Programmes

In an ideal proficiency programme, the samples would satisfy a number of important criteria. Firstly, they should be matrix matched to real specimens (in this case, human urine). Secondly, both parent compound and metabolites should be present in realistic concentrations. The exact concentrations of all analytes should be known. Further, samples should test both "normal" and "abnormal" situations. Finally, known or potential analytical problems should be challenged.

Like many ideal situations, it is impossible to meet all these criteria as a number of them are mutually exclusive. Most proficiency programmes use samples which fall into one of the following categories:

2.1.1 The Weighed In Control. In programmes using this type of control, an accurate amount of a compound is added to a blank matrix. Such controls have an "absolute" answer, and can be spiked to cover all sorts of realistic concentrations. False positive and false negative results are immediately obvious; single compounds or known combinations can be produced. In addition, sensitivity limits can be calculated. Unfortunately, this type of sample is usually not typical of those a laboratory receives in real life. In addition, many of the less common compounds, metabolites are either unavailable in sufficient quantity or are prohibitively expensive to purchase.

2.1.2 The Therapeutic Sample. Where a patient is on known medication, it is usually possible to collect sufficient urine for use as a control sample. Although concentrations are not known, all possible analytes will theoretically be known. The disadvantage of this type of sample is that it may be difficult to test odd combinations or non therapeutic situations.

2.1.3 The Emergency Sample. Where a patient has been admitted (usually for overdose), it is occasionally possible to get sufficient sample to provide a useful control. Such samples are realistic by their very nature. However there can be serious ethical problems involved, such as getting informed consent or the unnecessary use of a catheter to provide sufficient sample. It would be difficult to get sufficient samples of this nature to provide for a large, regular programme. Concentrations and analytes cannot be assigned with absolute confidence.

2.1.4 The Pooled Sample. Where a laboratory has access to a large number of urines, it is possible to pool samples with similar analytes. Since samples come from real patients, the pool will represent realistic levels of the major compounds. However, because patients may have different patterns of usage, there is potential for a very large pool of minor analytes to be present. To this extent, the pooled urine is less realistic than either the therapeutic or emergency sample. As is the case for the emergency sample above, it may not be possible to know the absolute concentration or identity of all compounds.

2.2. The AUSTOX programme.

The Toxicology Unit analyses over 60,000 urine samples per year. Samples cover most of the range of reasons for requesting drug testing, including hospital admissions, drug treatment programmes and medicolegal samples. Where analysis of a sample indicates the presence of a single significant compound, that sample is pooled with others containing the same compound and stored at -18°C. In an attempt to implement as many of the criteria of an ideal programme as possible, the AUSTOX programme uses a mix of real and weighed in samples. In any given year, approximately 30% of samples will be prepared from blank urine spiked with known concentrations of selected drugs. Allowing for the errors produced by the preparation of the samples (heating followed by a two stage freeze drying process), this allows some identification of false positive and negative samples as well as providing participants with valuable information about the sensitivity of their methodology. The remaining 70% of the samples are "real" samples, provided either by volunteers or from the stock of pooled urines. Each of the samples will have one or more major target drugs, chosen to challenge laboratories in their identification. They may be submitted in groups which test the abilities of laboratories to differentiate compounds which are structurally or analytically similar.

Because of the pooled nature of most of the samples, the programme functions by peer review. With the exception of the spiked samples, absolute answers do not exist. However, the monthly report for each sample is broken down according to methodology, and analytes are usually presumed to be present if detected by a significant number of laboratories using GC/MS. It is up to each laboratory to satisfy itself as to the correctness of its methodologies where it finds analytes not detected by others. A compound found by a single laboratory may be a result of misidentification, contamination or sample mixup. It may also occur when a laboratory is performing a unique analysis or has lower detection limits than others.

Similarly, when a laboratory does not find a particular analyte, it may indicate a deficiency in its detection techniques in either sensitivity or identification. On the other hand the drug may be one in which the laboratory has no interest and hence does not attempt to identify.

Like any proficiency programme, maximum benefit can be gained by the participants when samples are treated in the same manner as normal specimens. It is highly recommended that all samples be run as total "blind" samples. Unfortunately, it may not always be possible to arrange such a situation. However, in this programme, most laboratories which are unable run samples anonymously reported that they attempt to treat them as routinely as possible.

3 PROGRAMME RESULTS

Bearing in mind the difficulties in assessing the correctness of any individual result, it is still possible to examine some trends from the programme. Where a sample has been spiked, there can be an absolute identification of false positive and negative samples. Similarly, if the sample was taken from a volunteer on known medication, results are generally unambiguous. False negative results can often also be assigned, although to some extent this depends on the service offered by individual laboratories. It may also be possible to assign suspected false positives in consultation with a participating laboratory.

Given these limitations, comparing the results of the first two full years of the programme showed a heartening improvement in results. The false positive rate dropped by a mean of 18% (range 0 - 43%). Furthermore, the detection rate of true positives increased by 33%.

Of the 9 laboratories involved in the original programme, 8 were still involved in 1993. The results for these laboratories were compared with their 1983 results and additional improvements noted. The false positive rate dropped by a further 40%. The mean false positive rate among the original laboratories is approximately 10%. It should be noted that these laboratories report very few false positives of routine drugs of abuse, and that typical false positives are as a result of misidentification of relatively new therapeutic compounds. A further increase in the true positive detection rate of 8% has been demonstrated.

Analysis of the results of individual laboratories after they enter the programme show a similar pattern: there is a marked improvement in the both the false positive and false negative rate of most laboratories over an initial period of 12 to 18 months, followed by a slower but measurable period of improvement over time.

One important area where the programme has aided the improvement of individual laboratories is in the identification of contamination problems, poor sample tracking or unsatisfactory checking. Occasionally, laboratories have reported false positives for the same drug in all three samples, clearly indicating contamination somewhere in their procedures. Such problems are normally solved by the next month's cycle and have not recurred within an individual laboratory. Similarly, results for one sample are sometimes incorrectly transcribed onto the wrong report form. Again, correction is generally swift and permanent.

The programme also highlights a number of deficiencies in the way particular technologies are used and their results reported. For example, although only 15% of laboratories enrolled in the 1994 programme used immunoassays as their sole measure of detection, approximately 50% of identifiable false positives were reported by these laboratories. In particular, 80% of all Amphetamine false positives were attributed to immunoassays. While not unexpected, the results emphasise manufacturers' recommendations that positive results must be confirmed by an independent methodology.

A number of drugs or drug classes have been identified as presenting serious detection or identification problems for a significant proportion of those laboratories expecting to detect them. These include drugs or metabolites with serious abuse potential, including pethidine, monoacetyl morphine, and a number of the more potent benzodiazepines. Some commonly used therapeutic compounds can also present detection problems.

Clearly, any improvement in performance must be credited to the laboratories who have invested in better technology or sharpened their existing methodologies. The role of the AUSTOX programme has been to identify deficiencies in individual identification techniques, to provide an ongoing monitor that laboratory procedures meet a satisfactory level, and to provide a forum for education and exchange of views to its members.

Reduction of Errors in Laboratory Test Reports: Comparison of Continuous Quality Improvement Techniques with Laboratory Information System Techniques

Leslie Burnett and Judy Banning

DEPARTMENT OF CLINICAL CHEMISTRY, INSTITUTE OF CLINICAL PATHOLOGY AND MEDICAL RESEARCH, WESTMEAD HOSPITAL, WESTMEAD, NSW 2145, AUSTRALIA

1. ABSTRACT

We have compared two different strategies for minimising the number of addressing errors on clinical laboratory test reports over an extended period of time. One strategy, consisting of the application of continuous quality improvement (CQI) techniques involving manually implemented control charts, was successful in maintaining error rates at minimum levels for two years. An alternative strategy, consisting of implementation of an integrated laboratory information system with electronic data interchange of a hospital patient master database, was then implemented. We have found using CQI techniques resulted in error rates 16-fold lower than that achieved using fully computerised procedures.

2. INTRODUCTION

The issuing of a printed test report or chart is the usual process by which the final results of analysis are communicated from the testing laboratory to the requesting practitioner.

In an earlier article,[1] we have described how the systematic application of continuous quality improvement (CQI) techniques achieved a 17-fold reduction in the rate of incorrectly addressed test reports. This improvement in report quality was accompanied by faster turnaround time, no significant expenditure of funds, and no changes in staff, equipment, or in the laboratory information system.

The present study was designed to answer the following two additional questions:

1) Were the gains made in the original study able to be maintained for extended periods of time?
2) Could the same (or greater) gains also be made by computerisation of the work procedures?

3. METHODS

3.1 Description of laboratory

The Department of Clinical Chemistry, within the Institute of Clinical Pathology and Medical Research at Westmead Hospital, is a Department within one of Australia's largest medical testing laboratories. It provides comprehensive laboratory services to a 900-bed tertiary referral and university teaching hospital, to three district hospitals, and also provides reference laboratory services to other pathology laboratories in Sydney. It operates 24 hours a day.

3.2 Laboratory Information System

The Department's laboratory information system (LIS) is as described previously.[1] In brief, it is a 12-year old computer system that supported only the Clinical Chemistry Department's activities. It was unable to obtain information directly from Westmead Hospital's patient database. This necessitated all address information being manually transcribed and entered into the LIS from handwritten request forms.

In October 1993, this Departmental LIS was completely replaced by an organisation-wide LIS, the Cerner PathNet LIS.[2] The PathNet LIS was a larger and more complex LIS integrating results from all laboratory Departments within the Institute. The PathNet LIS was directly interfaced to the Westmead Hospital Patient Master Index database. This meant that information on the correct location for patients and the correct address for pathology reports could be obtained by the LIS directly by electronic data interchange.

3.3 Error Rates

Prior to September 1993, at the end of each month the Departmental LIS database was examined to determine the total number of Final Pathology Reports printed for the Clinical Chemistry Department, and the number of these Reports which did not have a valid "source code"; this was considered equivalent to not knowing the destination address.[1]

Conversion from the old LIS to the new occurred in October 1993. Data for this study was deliberately excluded for the period October to December 1993, to allow three months for the new LIS to "settle in".

From January to July 1994, the PathNet LIS database was examined in the following manner:

1) All accessions on which a Clinical Chemistry test had been requested were identified and counted; these were regarded as equivalent to the total number of Final Pathology Reports that would have been printed had the previous LIS still been in operation.
2) Each of these identified accessions was then examined to determine the number of these accessions that did not have a valid "client code"; this was considered equivalent to not knowing the destination address.

4. RESULTS

In our earlier study covering the period July to November 1991[1], we demonstrated how the systematic application of continuous quality improvement (CQI) tools[3] such as cause-and-effect diagrams, Pareto analysis and control charts, enabled us to reduce the number of wrongly addressed pathology reports 17-fold, from 1.33% each day to less than 0.08% each day. The procedures developed during that study were documented, appropriate staff training instituted, and these procedures were then implemented into daily laboratory operations.

Using a monthly Shewhart p control chart[4, 5] as a sentinel measure[6] we monitored the effectiveness of this solution over an extended period of two years, from November 1991 to September 1993 (Figure 1). We found that the original solution developed by the CQI team was sufficient to maintain the rate of wrongly addressed reports at a statistically stable[4, 5, 6] rate of 0.10% each day, which was equivalent to fewer than one wrongly addressed report each working day.

In October 1993, our laboratory's LIS was superseded by a new, organisation-wide LIS. Project planning and implementation of the LIS was undertaken by professional project managers employed either by the LIS vendor, or by our own organisation. The procedures developed by the CQI team to correctly enter client addresses were dismantled during this implementation of the new LIS because it was believed by the project managers that the new LIS could obtain this information electronically from the Hospital Patient Master Index database.

Using the same Shewhart *p* control chart, we measured the equivalent error rate for wrongly addressed pathology reports using the new LIS (Figure 1). Even with electronic data interchange from the hospital database, this error rate was now 1.62%, which was 16-fold higher than that achieved using manual implementation of process improvement techniques. The error rate was also found to fail tests of statistical stability.[4, 5, 6]

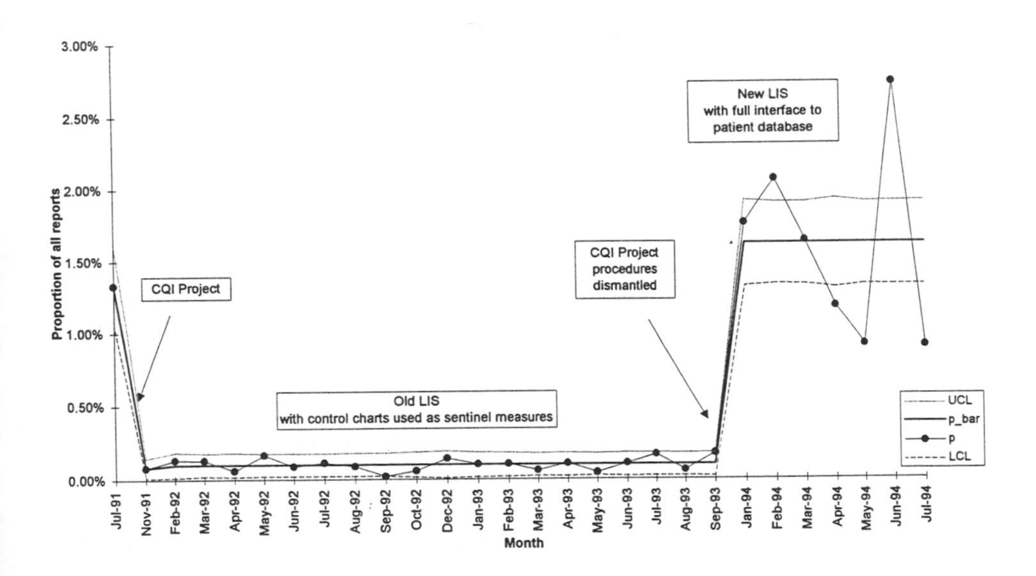

Figure 1. *Shewhart p control chart of the number of pathology reports each month that lacked an address. Each month, the proportion ("p") of all final pathology reports printed which lacked a valid address was charted. The average proportion ("p_bar") was calculated for each of three periods: prior to the CQI project, following the CQI project, and following conversion to the new LIS. Upper and Lower Control Limits (UCL, LCL) were calculated, and represent the 3σ limits within which approximately 99.7% of variation would be expected to occur by chance alone.*

5. DISCUSSION

The first part of this present study demonstrates that improvements made can be held over considerable periods, by:

1) using simple sentinel measurement techniques to
2) keep a watching brief on the process and to
3) alert the laboratory to any deterioration in process performance.

Our earlier study[1] demonstrated that process improvement using CQI techniques and tools could achieve a 17-fold reduction in error rates, without expenditure of funds and with a reduction in resource consumption. The actual error rate maintained over a two year period was 0.10%, which was not significantly different (using 3σ control limits) from the level of 0.08% achieved by the original process improvement team.

The second part of this present study demonstrated that computerisation of the process using the most modern and sophisticated tools available to our organisation resulted in a 16-fold deterioration in process performance. This represented a reversion to error rates comparable to that of pre-study performance.

Information collected by the original process improvement team[1] showed that only 0.1% of reports had root causes for reports being wrongly addressed which were directly attributable to a lack of information in the hospital database. The remainder of root causes consisted of a collection of unrelated procedures and details that the project team had to discover by cause-and-effect brainstorming, and the collection and analysis of data to identify the relevant factors. The lack of statistical stability shown in Figure 1 supports the interpretation that the likely causes of wrongly addressed reports from the new LIS will now lie in variation in procedures, either because these procedures have never been developed, or because of the inappropriateness of, or insufficient documentation of, procedures and training of staff in the relevant procedures.

It is a fact of life that errors will be made in even the best managed and operated laboratories. Quality Management philosophies and techniques allow one to measure, control and improve the level of quality of laboratory products and services. Laboratory Managers will often be faced with the question of whether quality problems can be addressed better by intensive process improvement project teams using CQI tools, or by investment in new technology and automation. The Quality Literature contains examples of generic warnings against reliance on the sole use of computers or automation to improve quality.[7] This study is a salient example of how investment in new technology within the laboratory will need to be integrated into the overall quality improvement process, or else those gains made will be rapidly lost.

Acknowledgments. We thank John Baulderstone for his assistance in extracting data from the Cerner PathNet LIS database, and Mark Mackay and Doug Chesher for their critical reading of the manuscript.

---

1. J. Banning, J. Brown, L. Hooper, J. Hamilton, J. Burnett and L. Burnett. Reduction of errors in laboratory test reports using continuous quality improvement techniques. *Clin. Lab. Management. Review* 1993, 7(5):424.

2. Cerner Clinical Information Systems, Version 304, Cerner Corporation, Kansas City, Missouri USA.
3. J. McConnell. 'The Seven Tools of TQC'. 3rd edition. Delaware Books, Dee Why NSW Australia. ISBN 0 9588424 0 4.
4. J. McConnell. 'Analysis and Control of Variation: Control Chart Techniques for TQC Practitioners'. 3rd revised edition, 1987. Delaware Books, Dee Why NSW Australia. ISBN 0 9588324 0.
5. E.L. Grant and R.S. Leavenworth. 'Statistical Quality Control'. 6th edition, 1988. McGraw-Hill, New York NY. ISBN 0-07-024117-1.
6. Western Electric Handbook Committee. 'Statistical Quality Control Handbook'. 2nd edition, Delmar Printing Co., Charlotte, NC: AT&T Technologies, 1958.
7. W.E. Deming. "Out of the Crisis'. 1986. Massachusetts Institute of Technology, Center for Advanced Engineering Study, Cambridge Mass USA. ISBN 0-911379-01-0.

TQM in a Government Laboratory: Salvation and Nightmare

E. M. Gibson

ENVIRONMENTAL CHEMISTRY UNIT, ENVIRONMENT PROTECTION AUTHORITY, GPO BOX 4395 QQ, MELBOURNE, VICTORIA 3001, AUSTRALIA

1 INTRODUCTION

The catchphrase "Total Quality Management" has been spoken for many years as the saviour of industrialised nations languishing economically due to poor productivity and management. The underlying philosophy of TQM has been defined as 'continuous improvement in the performance of all processes and the products and services that are the outcomes of those processes'[1]. The origins can be traced to the postwar reconstruction period in Japan using statistical techniques to control the quality of manufactured goods and know as Total Quality Control[2]. Other concepts developed subsequent to TQM are: world class manufacturing, world best practice, value adding management-manufacturing, zero inventories and Kaizen[3]. The common components of these concepts are based on:

- continuous improvement
- meeting or exceeding customer requirements
- employee involvement and empowerment
- management of variation
- provision of job security
- understanding suppliers processes

The concept is relatively easy to embrace and implement for a small organisation and a small budget, without consultants, providing there is a strong commitment from management and knowledge of the processes involved. The benefit to organisations is very real. However, the major barrier is ensuring continuity and acceptance by middle management. This paper's intent is to describe how the philosophy of TQM has been applied to a Government laboratory.

1.1 The Environmental Chemistry Unit

The Environmental Chemistry Unit (ECU) is a laboratory providing specialist analytical services within the Environment Protection Authority (EPA), State Government of Victoria. The EPA provided a suitable environment for the implementation of TQM by allowing the Units to be autonomous with their own budgets and business plans. ECU has 18 staff, two thirds are graduates with half of those having postgraduate qualifications and a budget of $900K. The core services (internal to the EPA only) are analysis of environmental media for prosecution and investigation purposes; stationary source emission testing; ambient air analysis; designing and/or auditing of environmental and industrial monitoring programs; 24 hour emergency response monitoring; consultants for air, water, waste and soil pollution issues; expert witnesses and selected research projects. Routine analysis are outsourced.

The laboratory is well equipped, for example, GC-FID/ECD/MS; purge and trap-GC; canister sampling GC-MS; HPLC-diode array/UV-VIS/; FTIR; graphite furnace AA. The laboratory is accredited under the National Association of Testing Authorities, Australia. Originally ECU was divided into two laboratories (with separate NATA accreditation) and the reporting structure shown below:

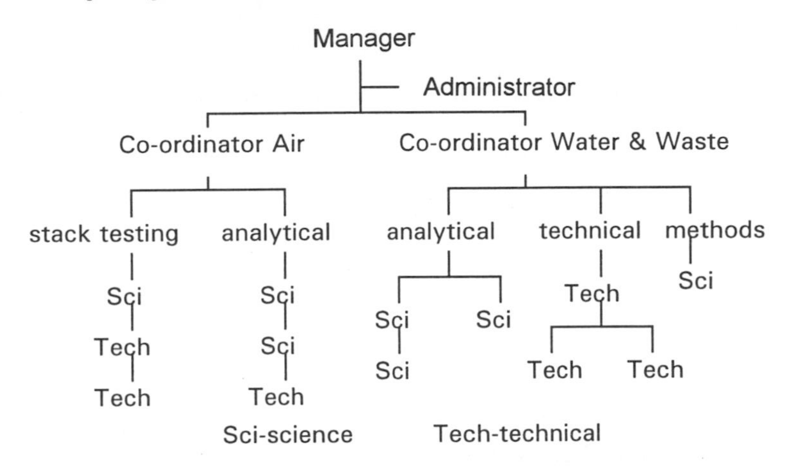

Figure 1 *Reporting structure of Environmental Chemistry Unit*

Accommodation was not ideal due to location of the Unit on several floors. The Air laboratory was located on the 1st floor, the Water and Waste laboratory on the 2nd floor and the manager on the 3rd floor. Other units within the EPA also shared this accommodation.

1.2 Audit of Environmental Chemistry

In 1993, an audit was conducted of the services provided. It demonstrated a lack of commitment to clients, low productivity, poor turnaround times whilst conducting research projects at the periphery of the EPA's mission. The audit was conducted prior to a manager of the two laboratories being appointed. A list of recommendations to be implemented was handed to the coordinators of the two laboratories. Little attempt had been made to implement these recommendations when the manager was appointed 6 months later.

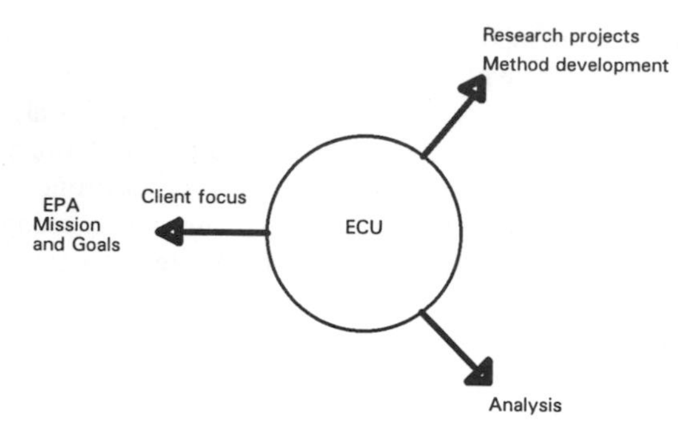

Figure 2 *Conflicting activities of ECU preventing a focussed approach to clients*

The other issues were:

- Little communication between areas, preventing the sharing of resources and skills.
- Low morale.
- Career prospects perceived as non existent and only through controlling people.
- An extremely hierarchal structure.
- Poor credibility within the EPA.

2. IMPLEMENTATION OF TQM

There is no one set way of implementing TQM: it will vary from organisation to organisation. The following will illustrate how the philosophy of TQM was used to improve the performance and working environment of the ECU. The term adopted for the process was Continual Improvement Process, (CIP).

2.1 Initial Stages of Implementation

The philosophy of continual improvement was communicated to the staff and continually reinforced. The structure of the ECU needed to be changed; however, this was difficult due to locations of staff. Whilst plans for relocation of the Unit were being put forward, there were major issues that needed to be resolved immediately in order for the Unit to progress.

2.1.1 Client Management. A team of people from both laboratories was set up to investigate client needs. The team consisted of 8 staff and formulated a questionnaire. Staff were assigned Operation Units to interview on a face to face basis. Previously, staff had not been involved in any formal contact with clients. Two major benefit of this exercise were the realisation by ECU staff that they had clients whose requirements had to be met and better communication between the staff and clients. Previously, ECU was seen very much as a black box: sample in at one end, report out at the other.

2.1.2 Benchmarking. Benchmarking was conducted against similar Government organisations in the States and Territories throughout Australia. This was mainly to identify how each organisation was set up with the varying constraints under which Government laboratories work and to identify the range of specialist analytical services offered.

Benchmarking was also done against commercial laboratories for areas such as turnaround times.

2.1.3 Removing barriers to getting on with the job. A team of staff directly involved with the analysis were put together for 8 weeks to determine why there was poor productivity with respect to analysis. The team brainstormed the causes of slow turnaround times. The causes were categorised and solutions put forward. Individual team members implemented the solutions where possible. A flow diagram for the progress of a sample through the laboratory was put together and the slow points identified. The information was communicated to the rest of the Unit.

2.1.4 Project Management. The projects undertaken varied from method development to major environmental studies. A project proposal system was put in place, where the project leader was identified as the person doing the bulk of the work (previously this role was taken by the coordinator). A project proposal proforma required justification, budgeting (including labour), a time plan and had to be approved by the manager. The project leader was responsible for completing the project and progress reporting using a proforma on a monthly basis. The highlights from each project together with bar charts for % on budget and % on schedule were collated for all the projects and included in the ECU monthly report which was discussed with staff.

2.1.5 Performance Planning and Assessment. Little attention had previously been paid to assessment with increments granted on a yearly basis almost automatically. With the advent of contracts and performance based awards on a corporate level flexibility was introduced. There was now an avenue for rewarding and promoting staff on the application of their expertise and judgement as opposed to previously where the more people being controlled was seen as the only way to be promoted, thereby losing valuable skills from the Unit. Staff were now formally assessed annually where outcomes with performance indicators are negotiated for the following year together with development planning. Informal feedback is provided throughout the year.

2.2 Continuous Improvement Process

CIP began in earnest with the combination of the laboratories. This was a major issue with the staff from the air laboratory. They agreed with the benefits of combining the two outlined to them but, in the beginning, did not believe it justified the move. The way it was conducted again was through the use of a team incorporating staff from both areas. The team had the role of implementing the move. This involved building of offices, rationalising instrumental equipment, adopting single systems such as ordering, sample registration, reporting. All staff were kept up-to-date with the progress and had input into such areas as seating and siting of equipment. The actual process of moving went very smoothly taking approximately 6 weeks, including the building of office space and removing a wall in the laboratory. There was some loss in productivity but not below previous performance levels.

2.2.1 Implementing a Team Based Structure. The staff welcomed the team based structure as shown below. Team leaders were assigned with the responsibility of assigning sampling/testing and ensuring staff cover. In the first instance, they were encouraged to meet in their teams, first thing, on a daily basis for about 10 minutes. It was fully accepted that reward and promotion could now be attained through application of expertise to meet ECU's goals and not only through the number of people under supervision.

Manager

Field testing team	Quality Management	Document Control	Organic team	Inorganic team

Figure 3 *Team based structure of ECU*

2.2.2 The Mission Statement. The following is the mission statement of the ECU:

The mission of ECU is to provide the EPA with high quality analytical, scientific and technical expertise to internal and external clients in order to support the research, policy and enforcement programs of the EPA. We are therefore committed to:

- Meeting the agreed scientific, financial and time requirements of all internal and external clients.
- Using world best practice and the principles of total quality in providing specialised analytical services, documentation and communication of information.
- Maintenance and continual improvement of a highly skilled expert and motivated work team.
- Use of safe and efficient work practice and processes.

3 SUMMARY

The principles of TQM were introduced 18 months ago. This resulted in removing barriers preventing improved performance; emphasising the importance of satisfied clients for survival; introducing a team based structure; and using performance assessment to reward outcomes. Initial improvements were seen after 2 months; major improvements after implementing a team based structure. This has led to increased productivity (250%), improved turnaround times (300%), relevant projects being completed, research projects funded through external grants, publications being produced and improving credibility within the EPA at all levels.

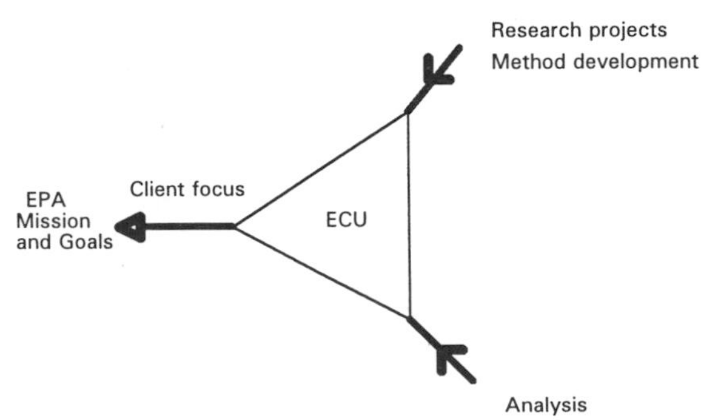

Figure 4 *Client focussed output of ECU*

The philosophy of TQM has to be continually emphasised and reiterated. It cannot stop because certain achievements have been made. Finally, an important point to highlight, is quality of the product was not an issue here with ECU being NATA accredited. Quality certification (ISO9000) or, most important for laboratories, quality accreditation (NATA) does not necessarily result in a high performing organisation despite producing a quality product. This is due to the lack of client focus which is only embraced in the philosophy of TQM.

References

1. B. Irwin, 'Quality Management', Proceedings of the Sixth National Logistics Conference, Sydney, 1990.
2. W.E. Deming, Quality Productivity and Competitive Position, Massachusetts Institute of Technology Press, Cambridge, Massachusetts, 1982.
3. P. Gilmour, 'Operations Management in Australia', Longman Cheshire, Melbourne, 1991.

Implementation of Quality Programs in Multifunctional Laboratories

J. W. Hosking

CHEMISTRY CENTRE (WA), 125 HAY ST., EAST PERTH, WESTERN AUSTRALIA, 6004, AUSTRALIA

1 INTRODUCTION

The Chemistry Centre (WA) and its predecessor, the Western Australian Government Chemical Laboratories, have provided analytical, chemical and related scientific services to some 30 government agencies, to numerous industries and to the general public for approximately 100 years. These services support diverse government programs and industries involved with agriculture, the environment, food, forensic science, occupational and public health, materials, mineral exploration, mining, mineral processing and racing.

The primary activity and strength of the Chemistry Centre has been, and is, analytical chemistry. The 80 professionally and 40 technically qualified staff of the Chemistry Centre have expertise in inorganic and organic analysis of most types of samples. Their efforts are supported with an extensive range of sophisticated equipment. The major output has been analytical test results. However there has been an increasing requirement for reports to include an interpretation of the numerical result. In recent years there has also been an increasing demand for the Chemistry Centre to use its analytical expertise to solve problems, and undertake investigations and research and development projects for its clients.

The Chemistry Centre generally does not analyse industry generated exploration geochemical samples, mine grade samples nor farm soil samples; this type of work generates significant business for many other laboratories. Instead the Chemistry Centre specialises, for example, in the analysis of whole rock geological samples, mineral processing samples and samples from statistically controlled agricultural field trials.

The mix of work has changed. Client requirements, commercialisation of the Chemistry Centre and competitive tendering for government business have contributed to these changes. The changes are schematically shown in Figure 1.

Routine analytical work includes inorganic and organic major, minor and trace analysis of geological, soil, plant and water samples and of a range of materials using well established methodology. Advanced analytical work includes, for example, inorganic and organic ultra trace analysis and the analysis of micro forensic samples. It also includes the analysis of toxins in foods, of antinutritional components in plant materials and drugs in biological materials. Some examples of problem solving work which utilise analytical expertise include the identification of unknown contaminants in foods, work place odours and environmental spills. Examples of investigations where analytical expertise has been

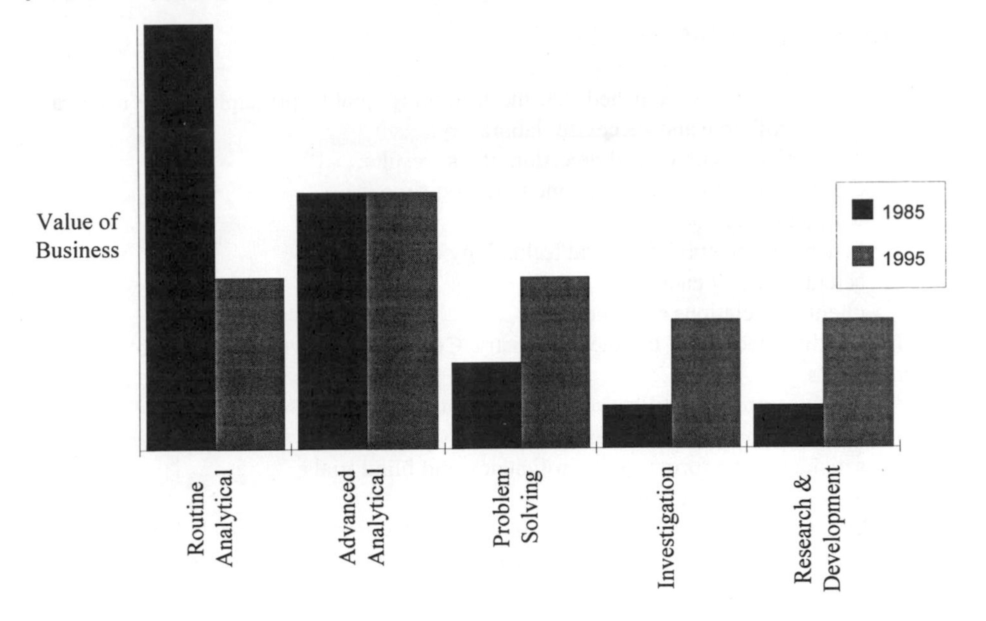

Figure 1 *Change in work type*

crucial to their successful completion include the characterisation of fretting mortar, the isolation, characterisation and synthesis of components of natural products, establishing guidelines for the avoidance of corrosion of mine rock bolts and forensic gold fingerprinting. Funded research projects which have developed from analytical expertise include development of methodology for the identification of forensic fibres and glass fragments, and arsenic speciation in occupational biological samples. Collaborative research projects continue to be undertaken with other government agencies, universities and CSIRO. The analytical and associated expertise of the Chemistry Centre complement other research skills; examples include the minimisation of nutrient export from agricultural activities, the destruction of waste chemicals, the identification of compounds causing hydrophobicity in sandy soils and the Enzyme-Linked Immunosorbent Assay techniques for the determination of drugs and lupin alkaloids.

The Chemistry Centre has been accreditated by the National Association of Testing Authorities (NATA) for nearly 50 years. The range of accreditated chemical tests has been significantly extended during this time. This history of involvement in formal quality systems demonstrates strong commitment from management and the scientific staff. Their understanding and familiarity of the principles of quality programs has developed as the requirements for accreditation have been regularly upgraded, particularly over the past ten years.

The implementation of quality systems and quality management principles to all Chemistry Centre operations was a high priority. This was needed to ensure that the Chemistry Centre continued as a profitable business recognised for excellence in the application of science and in satisfying client expectations.

2 QUALITY REQUIREMENTS

The Chemistry Centre established that the following quality principles were necessary to continue as a profitable and successful laboratory:

- appropriate accuracy and precision of test results;
- the right result is always obtained the first time;
- a target of zero defects;
- benchmark methodology and technology are used;
- scientific excellence;
- client expectations are exceeded.

It was thus necessary for the Chemistry Centre to implement the following quality requirements:

- quality systems certification for all operations;
- accreditation for regular tests;
- satisfactory performance in proficiency and blind trials;
- total quality management.

Quality systems certification (ISO 9000) covers the first requirement. Test accreditation (ISO Guide 25) covers the second requirement. NATA arranges extensive proficiency trials; however these only cover the minority of test work undertaken in the Chemistry Centre and therefore there was a need to participate in additional trials. There was also a need to integrate the principal elements of total quality management into the Chemistry Centre's quality systems.

The broad coverage of the ISO 9000 standard challenges the organisation seeking certification to ensure that it implements the most appropriate quality systems. The Chemistry Centre chose to implement requirements equivalent to those of ISO Guide 25.

3 QUALITY SYSTEMS

ISO/IEC Guide 25 (SAA HB 18.25 - 1991) states "Laboratories meeting the requirements of this Guide comply, for calibration and testing activities, with the relevant requirements of the ISO 9000 series of standards including those described in ISO 9002 when they are acting as suppliers producing calibration and test results."[1] The routine analytical work, some of the advanced analytical work and some analytical work supporting other areas shown in Figure 1 is undertaken on a regular basis and therefore able to be covered by ISO Guide 25. The one-off or infrequent analytical determinations, interpretations, investigations and research and development projects are not covered. These latter functions are however covered by the design/development ISO 9001 standard. Many of the financial and clerical functions are also not covered by Guide 25 but are covered in ISO 9000 standard.

Commonwealth and state governments and major companies have quality system requirements for service providers. This has certainly increased the number of businesses which have recently obtained quality system certification. However ISO Guide 25 test accreditation receives a lower level of recognition in the Western Australian Government Quality Assurance Policy; ISO 9000 certification is required for 'critical' purchases whilst test accreditation is required for 'non-critical' purchases.[2]

The Chemistry Centre chose to extend ISO Guide 25 accreditation to cover as wide a range of tests as possible and implement quality systems certification to ISO 9001. The quality systems established for ISO 9001 were modelled on ISO Guide 25 and on considerable project management experience.

The scientific and technical staff have a well deserved reputation for the quality and scientific excellence of their work. The quality systems developed to support this reputation resulted from the need to:

- be at least nationally recognised in all scientific areas;
- maintain NATA test accreditation;
- meet changing client needs;
- provide forensic, regulatory and other court or enquiry expert evidence;
- undertake umpire analyses and be a state reference centre;
- fulfil a range of statutory requirements;
- meet the obligations of research funding bodies;
- upgrade staff support systems.

Consequently quality systems have been developed and used in all analytical areas for many years. These have been modelled on NATA test accreditation requirements but have also recognised the above obligations. The elements of the systems included the following:

- unique sample registration;
- permanent laboratory work books, records and copies of reports;
- duplicates and control samples;
- documented methods;
- checking of all results and reports by experienced staff.

Quality manuals, staff training competencies and audits have been a requirement of more recent years.

It was observed that the metallurgical and mineralogical areas had much more difficulty accepting the need for even basic quality systems including the documentation of quality manuals and procedures; this is also probably reflected by the lack of standard methods for much of the test work in these areas. The financial, clerical and computing - information technology support staff were also not used to the concepts of quality systems; however the staff in these areas accepted the challenge and prepared and use procedure and area quality manuals.

With a limited range of NATA organised proficiency trials it is necessary for the Chemistry Centre to participate in trials arranged by other organisations. The Chemistry Centre has organised some trials but like most laboratories can no longer afford to finance such trials. Consequently there is a need for more collaborative proficiency trials. A limited number of blind trials are arranged through clients but more are needed.

Audits are often seen to be a nuisance. Experience has confirmed that the first is the worst. The benefits are significant and they are crucial to the improvement of laboratory procedures, processes and quality systems. Experience has also shown that audit independence and management commitment are essential. Likewise check lists prove valuable; but the fresh 'eyes' of an experienced laboratory person provide the most benefit.

4 TOTAL QUALITY MANAGEMENT

The basic elements of total quality management include quality systems, continuous improvement and client focus. Many of the scientific staff have achieved a good understanding of the basic elements of total quality management without the need for formal training. They are innovative in method development, in the implementation of new technology and instrumentation and in the development of upgraded quality systems. They recognise the value of lateral thinking and the value of utilising ideas from other disciplines. They successfully utilise teams to solve scientific problems. Many have implemented very significant efficiency improvements and can rightly claim to have established the "industry" benchmark. However it was recognised that the concepts of continuous improvement could benefit all areas. Likewise although good client contacts had been maintained by some staff in general significant improvements continue to be made in this area. A consultant was used to assist with staff training and the implementation of these programs.

4.1 Continuous Improvement

The success of the continuous improvement program varied from area to area. A range of 'barriers' to improvement were identified during the training sessions. Some of these 'barriers' clearly indicated communication problems as systems were already in place which should have resolved any perceived or real 'barrier'. There was some reluctance to include all operations in the continuous improvement process. Some staff were not comfortable in addressing improvements in areas where they were not directly involved even though they had an impact on that area, for example sample preparation or invoicing. Some staff were also reluctant to accept the need for change and for the need to make comparisons and to benchmark; their attitude was 'if the result is accurate and precise why change ? '. Some managers appeared to see the process as a diminution of their authority and therefore did not fully support the program. Some groups have however had significant successes while others have been frustrated by perceived or real barriers.

The complaints process of ISO Guide 25 and the corrective action processes of the quality systems ISO 9001 standard were integrated with the continuous improvement program through the Chemistry Centre's Corrective Action Request Form. This form handles problems, system failures or improvement ideas to ensure that they are handled and not ignored. These forms are initiated by any staff member in response to any of the following:

- safety problem;
- supplier, QC or internal system failure;
- audit or management review;
- customer complaint;
- idea for improvement.

The continuous improvement processes are monitored by the Executive and support given at every opportunity to ensure maximum benefits.

4.2 Client Focus

A recent independent survey of Chemistry Centre clients confirmed high client satisfaction except for the area of turn-around. Clearly sample turn-around needed a major effort. However all other areas of the operation also needed reviewing to improve client satisfaction. The Chemistry Centre like any business also needed to expand its client base. It was thus necessary to implement client focus principles to all staff in the Chemistry Centre and to feed improvement ideas into the continuous improvement programs.

For many scientific staff this change provided many challenges. For example, the concept that clients receive the most cost effective solution to their problems even when this did not allow the scientists to continue until all answers had been obtained. It was also necessary, particularly for the problem solving and investigative work, for the client requirements and expectations to be documented and agreed to prior to the commencement of the work. Where the outcomes cannot be forecast the quality systems require the work to progress as a series of small projects.

Internal clients were often not well treated. Many problems and investigations require inputs from staff from other areas; in some cases this work was not given the required priority. Financial and clerical staff had difficulty accepting that their role was to support the laboratory staff; their natural instinct was to set the rules instead of concentrating on providing excellent service.

Appropriate software has been installed to record client contacts, and to ensure that regular contact is maintained, follow ups occur and that client leads are followed up.

5 CONCLUSION

The Chemistry Centre like most analytical laboratories has a history of active involvement in quality systems. Even so it has required considerable effort from staff to meet the challenges of the quality changes of the last decade. These challenges will continue with continuous improvement programs giving rise to quality system upgrades and other improvements in all areas of the organisation.

As stated in the AS/NZS ISO 9000.1:1994 'It is not the purpose of these International Standards to enforce uniformity of quality systems. Needs of organizations vary. The design and implementation of a quality system must necessarily be influenced by the particular objectives, products and processes, and specific practices of the organization.'[3] However there is a challenge also for the certifying and accrediting bodies to rationalise the guidelines and requirements yet maintain a consistent and credible system for analytical laboratories which will be nationally and preferably internationally recognised. If this is not achieved customers and potential customers will be even more confused.

References

1. Guide 25 - General requirements for the competence of calibration and testing laboratories, SAA HB 18.25 -1991, Standards Australia, Homebush, 1991, p. 5.
2. Quality Assurance Policy: Guidelines for Suppliers, State Supply Commission of Western Australia, 1993, p. 3.
3. Australian/New Zealand Standard, Quality management and quality assurance standards Part 1: Guidelines for selection and use, AS/NZS ISO 9000.1:1994, Standards Australia, Homebush, 1994, p.vii.

Method Validation: An Essential Element in Quality Assurance

I. R. Juniper

NATA AUSTRALIA, 7 LEEDS STREET, RHODES, NSW 2138, AUSTRALIA

What do we mean by 'method validation'? Validation is the process of establishing the suitability of a method for authorisation for use in the laboratory. Its purpose is to determine the applicability of a test method to a particular analysis of a specified sample type and to identify any limitations to its use.

We also need to distinguish between 'validation' and 'verification'. Verification is the quality assurance procedure whereby a laboratory regularly analyses reference materials, QA Samples and spiked samples, and carries out blank determinations as a means of confirming the validity and reliability of routine sample analyses.

Whether a recognised standard procedure is used, an existing method is modified to meet special requirements, or an entirely new method is developed, it is necessary to validate it. Laboratory studies must establish that its performance characteristics meet the requirements for the intended analytical operations.

1 DOCUMENTATION OF THE TEST METHOD

Before attempting to validate a method, it must first be written in such a way that the instructions which it contains are complete, unambiguous, clearly understandable, and can be followed without difficulty by a competent analyst. It should be written in a format similar to that specified in Australian Standard AS2929 Test Methods - Guide to the format, style and content, or ISO 78-2 Chemistry - layout for Standards Part 2: Methods of chemical analysis, and should be a controlled document from the moment of its inception.

A documented test method which is to be approved for use should include a statement of the scope and range of applicability of the method, and a summary of the validation data, which should include the recovery and coefficient of variation over the concentration range over which the method is applicable. The method should also specify the criteria for acceptance of test results, and the action to be taken in the event of nonconforming results being obtained.

2 INTENDED APPLICATION OF THE TEST METHOD

In order to decide upon the means to be used to validate a method, the purpose to which it will be put must be clearly defined and a number of factors considered:

- What kinds of sample matrix are to be analysed?
- What are the analytes to be determined, and what are their expected concentration levels?
- Are there likely to be any interfering substances in the samples to which it will be applied?
- Are there any specific legislative or regulatory requirements, such as action levels or limits of reporting, to be met by the analyses?
- What analytical instrumentation and analytical skills are available?
- Is it to be a screening procedure capable of detecting and identifying a number of compounds with similar physical or chemical characteristics, or is it to be specific for a given analyte?
- What level of detection is required?
- How robust should the method be? Is it intended for use by a skilled analyst or by a technical assistant?
- How much is the client willing to pay for an analysis?

Once these matters have been decided upon, a selection of the analytical parameters to be evaluated can be made. These parameters are:

- accuracy and precision (repeatability and reproducibility),
- selectivity (specificity),
- limits of detection and reporting,
- linearity,
- range of applicability,
- effects of substances known or likely to interfere with the determination which may be present in the sample,
- effects due to sample matrix,
- recovery of analytes (as a function of concentration).

While it may not be necessary to determine all of them for each test method, they should all be given consideration and the reasons for not determining any one of them documented. For example, if a determination is to be carried out for an analyte present at a relatively high concentration, the sensitivity of the method may not be as critical a factor as the need to establish linearity of response and to ensure that sufficient reactant is always present. Conversely, if an analyte is to be determined at very low concentrations, the minimising of interferences assumes greater importance.

3 PROCEDURES FOR EVALUATING AND VALIDATING TEST METHODS

Two IUPAC papers will be found of use when planning and conducting method validation studies. A *Protocol for the Design, Conduct and Interpretation of Collaborative Studies* was prepared by William Horwitz following the IUPAC Workshop on the Harmonisation of Collaborative Analytical Studies held in Geneva in May 1987. Following the second IUPAC Workshop held in Washington in April 1989, harmonised protocols for the adoption of standardised analytical methods, and for the presentation of their performance characteristics, were prepared for publication. These protocols were subsequently published in Pure and Applied Chemistry in 1988 and 1990.[1,2] The second set outlines the criteria to be considered before adopting a method for publication as a standard method, and the aspects of analytical quality control which are to be associated with the texts of standard methods, together with recommended formats for drafting of essential information on analytical quality control.

The validation study should be carried out by an experienced analyst in order to avoid errors due to inexperience since the purpose of the study is to evaluate errors and possible bias in the method itself. As a rough guide, a minimum of five analyses should be carried out under repeatability conditions, but it is recommended that at least ten analyses be carried out when determining the precision and accuracy of the method.

The procedures available for the investigation of these analytical parameters can be summarised as follows:

Table 1

Procedure	Parameters to be investigated	
Replicate analysis of samples	Repeatability	
Analysis of blanks	Interferences	Limit of detection
Analysis of reference materials	Accuracy	Repeatability
Analysis of spiked samples	Recovery Interferences	Repeatability Range of applicability Limit of detection Limit of reporting
Analysis of Standards	Accuracy Repeatability Linearity	Limit of detection Limit of reporting
Analysis by different operators	Reproducibility (ruggedness)	
Interlaboratory studies	Reproducibility	Accuracy

Before discussing how these parameters are to be evaluated, it is necessary to recall their definitions.

3.1 Precision and Accuracy

Every measurement made has an uncertainty associated with it. In order to estimate the uncertainty of measurement it is necessary to know the variability that occurs when repeated measurements are performed (precision), and the degree of departure of the measurement from the accepted value (accuracy).

Note: 'Imprecision' and 'inaccuracy' are numerical expressions of precision and accuracy.

3.2 Accuracy

Accuracy is the closeness to the true value of the test results obtained by the method. In other words, it is a measure of the exactness of the analytical method.

3.3 Bias

Bias is the systematic component of the error of a measuring instrument or a test procedure. Procedures can be found in text books of statistics for evaluating method bias using tests on the mean (t-test), comparison of two methods by least-squares fitting, comparison of the precision of different methods (F-test), and the chi-square test for comparing experimental results with expected or theoretical results.

3.4 Precision

Precision is a general term for the variability between repeated tests. Two measures of precision, termed repeatability and reproducibility, have been found necessary and, for many practical purposes, sufficient, for describing the variability of a test method.

3.5 Repeatability

Repeatability refers to tests performed under conditions that are as constant as possible, with the tests being performed during a short interval of time in one laboratory by one operator using the same equipment.

The determination of repeatability under such limitations has been criticised by many authors, including Horwitz, who assert that determinations of repeatability should be conducted over a period of, for example, several days, in order to take into account the variability which would be likely to occur over the time frame of the intended application of the test method. They consider that consecutive measurements can exhibit only minimal effects of variability and cannot demonstrate the systematic errors likely to appear over a longer period of time.

3.6 Reproducibility

Reproducibility refers to tests carried out in different laboratories by different operators using different equipment.

An obvious means of determining the repeatability and accuracy of a method is to analyse a standard or reference material for which the concentration of analyte is known with high accuracy and precision. The difference between the known true value and the mean of replicate determinations with the test method is due to the sum of the method bias and random errors.

This procedure suffers from the disadvantage that the result is only valid for the particular reference material used, and often no suitable reference material is available. If such is the case, the method being investigated should be compared with an existing or 'reference' method, for which it is often assumed that there is no method or laboratory bias. Unfortunately, this is not always true, and the final evaluation of a method should preferably be carried out by conducting an interlaboratory study.

When comparing different methods of analysis, the possibility exists of a procedural or instrument bias resulting from causes such as different sample treatment, different extraction and cleanup procedures, and the use of different methods of detection and quantitation. Ideally the results of both methods should be completely correlated, and statistical tests applied to determine whether the differences observed are significant.

The determination of the reproducibility of test results goes beyond the basic validation of a test method by a laboratory. It requires comparison of the results of analysis obtained by different analysts using the same procedure but with different instrumentation in different laboratories and at different times. This is best accomplished by regular participation in interlaboratory studies and proficiency studies.

3.7 Sensitivity, Limits of Detection and Reporting

The sensitivity of an analytical procedure can be defined as the slope of the calibration function $y = f(x)$. Sensitivities are seldom constant over large concentration ranges and are therefore only meaningful when concentrations or concentration ranges are specified. Often the lower boundary will be the detection limit.

The limit of detection is the lowest concentration of analyte in a sample that can be detected, but not necessarily quantitated, under the stated conditions of test. The limit of quantitation, or limit of reporting, is the lowest concentration of an analyte that can be determined with acceptable precision (repeatability) and accuracy under the stated conditions of test.

The limit of detection and the limit of reporting can be determined by analysis of replicate portions of a sample matrix to which have been added known concentrations of analyte. The detector response is then plotted against the concentration of added analyte. The imprecision (relative standard deviation or coefficient of variation) should be determined at concentrations increasing from the limit of detection until a concentration range is established over which the imprecision is acceptable.

By this method, the minimum level at which the analyte can be reliably detected is established. A signal-to-noise ratio of 3:1 for instrumental methods is generally acceptable, but some analysts calculate the limit of detection by determining the standard deviation of the background noise obtained by analysis of blank samples and multiplying this by a factor of three. The limit of reporting is frequently taken to be ten times the limit of detection. (For a discussion of the estimation of detection limits, reference should be made to Detection in Analytical Chemistry, ACS Symposium Series 361.[3])

3.8 Linearity and Range

The linearity of an analytical method is its ability to produce test results which are proportional to the concentration of analyte in samples within a given concentration range, either directly, or by means of a well-defined mathematical transformation.

The linearity of a method may be obtained by graphically plotting the test results as a function of analyte concentration. It is usually determined by calculation of a regression line using the method of least squares of test results against analyte concentrations. The slope of the regression line and its variance provide a mathematical measure of linearity, and the y-intercept is a measure of the potential method bias.

The range is the interval between the upper and lower levels of concentration of analyte that have been demonstrated to be capable of determination with the required precision, accuracy and linearity under the stated conditions of test.

3.9 Recovery

Recovery, or the proportion of an analyte present in a sample matrix which is capable of determination by a test method, can be determined by analysing samples to which have been added known amounts of analyte. This procedure has the limitation that the added analyte is not necessarily in the same form as that naturally present in the sample matrix and, because of its more ready availability, may give unduly optimistic recovery figures. In other words, recovery determinations give no indication of the accuracy of the unspiked sample result.

The analyst also needs to be aware that, in addition to the possibility that recovery of added analyte may be quite different from recovery of naturally present analyte, the recovery may also be dependent on the concentration of analyte present. It is therefore necessary to determine recoveries at a number of analyte concentrations ranging from near the limit of detection to near the upper limit of applicability of the method.

In USEPA methods for determining organic contaminants the recovery of a surrogate standard, that is, a reference compound which is similar in its analytical behaviour to the analytes under investigation, and which is added to the sample before analysis is begun, is used to determine the overall recovery of the test method. This procedure does not, however, overcome the difficulty of non-representative recovery.

Another approach to a solution of this problem was suggested by Dabecka and Hayward[4] in a paper presented at the Fifth International Symposium on the Harmonisation of Internal Quality Assurance Schemes for Analytical Laboratories held in Washington in 1993. They proposed a sample weight test, in which two different quantities of sample are analysed, one being twice the quantity of the other. If the same analyte concentration is obtained for the two sample quantities, then the result can be considered accurate. If different concentrations are obtained, then the cause should be investigated. This approach is, however, subject to limitations which may be imposed by a low analyte concentration, practical quantitation limits and the availability of sufficient sample for analysis.

A procedure not subject to this particular limitation is to analyse reference materials for which a certified value or a consensus value for an analyte has been determined. However, the test result obtained may differ from the consensus value because of differences in methodology. A further drawback is that the test result relates only to the reference material matrix, which in most cases is not identical with the sample matrix.

3.10 Robustness

When validating a test method for use in a laboratory, the robustness of the method, or its capacity to provide results of acceptable reliability in the hands of different operators, is best determined by replicate analyses performed by different operators over a period of at least several days, in order to include as many potential sources of error as possible.

Evaluation of a test method for acceptance as a method for general use by laboratories is much more demanding and requires the performance of an interlaboratory study involving at least six laboratories and the analysis of a number of appropriate sample matrices, both natural and spiked with the analytes of interest at a number of concentration levels.

As a further means of attempting to ensure comparability of test results, it may be necessary to provide each participant laboratory with standard mixtures of analytes at known concentrations, and to conduct a series of studies involving the analysis of solutions of analytes, sample extracts, and finally the sample matrix itself.

4 METHOD OPTIMISATION AND QUALITY ASSURANCE

During the course of validating a test method and determining its applicability and limitations the analyst will necessarily establish the analytical conditions which give optimum selectivity, sensitivity and recovery of the analyte. Interfering compounds may require development of a cleanup procedure or selection of a different means of detection or a different method of quantitation.

All investigations and conclusions should be recorded and documented in detail or much of the value of the work of validation will be irretrievably lost. There are few things more frustrating to a developmental analyst than to know that certain work was previously done but not to be able to find a record of it.

The laboratory should establish a quality assurance program in which blank determinations and the analysis of reference materials and/or check samples and spiked samples are carried out when samples are analysed. A typical quality assurance program might include the analysis of a blank, spiked sample and reference/check sample with every batch of not more than 12 samples, but the frequency with which such checks should be carried out is also dependent upon the technique of analysis and the robustness of the method.

Once the final details of the method have been established, the mean and standard deviation of the results of analysis over time of reference materials, check samples and spiked samples should be calculated and used to produce control charts.

Each successive result of analysis of a blank, spiked sample and reference/control sample should be plotted for comparison with a central line representing the mean, and with warning and action limits at 2 and 3 standard deviations respectively above and below the mean. Ninety-five percent of the results should lie within the warning limits. Any result outside the action limits, or two successive results outside the warning limits, indicates that the system is out of control and should be investigated. The results obtained in these batches are therefore suspect and the samples should be reanalysed.

A documented action plan should be implemented when analytical results fail to meet acceptance criteria, and records should indicate what corrective action has been taken in such an event.

5 REFERENCES

1. W. A. Horwitz, 'Protocol for the design, conduct and interpretation of collaborative studies', *Pure & Appl. Chem.*, 1988, Vol. 60, No. 6, pp 855-864.
2. W. D. Pocklington, 'Harmonized protocols for the adoption of standardized analytical methods and for the presentation of their performance characteristics', *Pure & Appl. Chem.*, 1990, Vol. 62, No. 1, pp 149-162.
3. L. A. Currie, 'Detection in Analytical Chemistry', ACS Symposium, Series 361, 1989.
4. R. W. Dobecka & S. Hayward, 'Missing Aspects in Quality Control', *Quality Assurance for Analytical Laboratories*, ed. M. Parkany, R.S.C. 1993.

The Promise of Quality, The Promise of Change

Daniel Kwok

CROSBY ASSOCIATES, LEVEL 8, 2 ELIZABETH PLAZA, NORTH SYDNEY, NSW 2066, AUSTRALIA

Clearly the last ten years in Australia saw a monumental expenditure of human and management resource in the pursuit of Quality Management. Once the province of the quality control and assurance professions, Quality Management became a fixture in general management. Concretised as "TQM" and given stature through national bodies and annual awards, Quality Management assumed a natural slot in the chief executive's agenda.

Has it delivered? Anecdotal evidence would suggest many organisations that throw considerable weight behind a quality initiative are not pleased with the results. They see a faltering of enthusiasm, a diminishing of returns, a returning to the way they were. They see the promise of a competitive edge devoured by bureaucracy and rituals. The attractive promise of a trouble-free business life fades.

Some move on to supplementary programmes such as business process re-engineering (BPR), world-class manufacturing (WCM) and benchmarking. Each of these is sought for its rejuvenation power, "to put us back on track". Others find succour in chasing accreditation with standards organisations or national quality awards both of which infer blessings from officialdom. This last statement often upsets people as many careers have been re-positioned to assist organisations achieve these pursuits.

In the seventies, many Australian companies were drawn to the QCC movement. This was almost a total waste of time. Replicating the most visible aspect of the Japanese success story in the hope of replicating their success is as nonsensical as a fresh immigrant claiming to be Australian after having mastered the art of a garden barbecue.

The truth is that programmes and campaigns do not change anything. The flurry of wall-to-wall training, posters and slogans, new tools, new words, power speeches, teams with heroic names, trophies, certificates - they all come to nought when pursued without the serious intent of successfully renovating the management lifestyle and day-to-day practices of the organisation.

The pursuit of Quality Management is the creation of an organisation that is capable of successfully tackling change.

Change is a constant since the beginning of time. Organisations may change by choice or by default. One gives you the potential of control. The other sees you fluttering in the wind.

Quality Management is not high-class problem solving. It is not a menu of improvement projects. It is not the creation of statistical freaks. It is not the proliferation of teams up and down the organisation (rather redundant when one considers that an organisation already consists of a multitude of natural work teams).

What all organisations, whether commercial or not, must consider is its ability to define accurately the requirements of those who receive their outputs - the customers. This is the first fundamental for success.

Following this is the ability to translate these requirements into the various work processes needed to achieve the outputs. Then comes the ability to make certain that all requirements can be met the first time round, and on time. This will guarantee delivering to the customer what was promised at the least cost.

Systems may be developed to make certain that requirements are defined and met. But systems are operated by people and people make mistakes. Where mistake-proofing devices are created, they should be installed. Other than that, a universally clear performance standard of Zero Defects should be role modelled and installed as a mind-set throughout the organisation. Philip Crosby, author of *"Quality Is Free"* and *"Quality Without Tears"*, sees errors or non-conformances caused frequently by lack of attention rather than lack of knowledge. He writes "Lack of attention is created when we assume that error is inevitable.....it is a state of mind......an attitude problem that must be changed by the individual". This change is necessarily a top-down culture change led by a sincere leadership adamant at making it happen. People calibrate to their leader's tolerance of errors.

All these would be easier to manage if our customer's requirements remain static. Of course they do not. Our customers' needs change constantly. Technology offers us new ways of doing things. Our stakeholders demand better and better performance. Government imposes new rules. Competitors up the ante as they plot daily to take our business away. Our market has cycles. Our workforce changes. Society develops and demands. The earth quakes.

We need an organisation that is able to anticipate emerging trends, new requirements and quickly swing its resources to meet them. Organisations must be ever-crouching ready to pounce at running targets. In less poetic terms, it must be willing to manage change. Change may mean an incremental improvement in a particular work process. Or it may mean severely re-engineering a key business process. It may employ SPC techniques. Teams may be involved. Or change may be executed by an urgent order. All these are the tools of change.

And what organisational leaders must focus on is not the tools, but the grander, more profitable objective of creating "changeability" in their domain. This requires transformation of the organisation into one that allows for creativity to flourish, for brave new ideas to enjoy sufficient passage to the right ears, for courage to destroy spent traditions and install new practices and beliefs.

Before we see this as a suggestion to slide into blue-sky oblivion, I must add that an organisation that manages change successfully needs to have the skills and systems to identify, plan, install and sustain specific changes. These do not come from rocket science. They are simple and common-sensical and the subject of another paper.

Quality Audits in a Forensic Science Organisation

Gavan J. Canavan and Michael J. Liddy

VICTORIA FORENSIC SCIENCE CENTRE, VICTORIA POLICE, FORENSIC DRIVE, MACLEOD, VICTORIA 3085, AUSTRALIA

1 INTRODUCTION

Over several years, the State Forensic Science Laboratory (now the Victoria Forensic Science Centre, VFSC) developed and implemented a total quality management system for the whole of the Laboratory's operations. The total quality management system was based on the relevant international standard[2] as well as the requirements of ASCLD-LAB, the American Society of Crime Laboratory Directors - Laboratory Accreditation Board[3].

The aims of this paper are to explain the need for internal quality audits as part of a quality management system, to detail the key components of a quality audit program, and to discuss the implementation of the quality audit program within our organisation.

2 QUALITY SYSTEM

Our total quality management system covers all aspects of the Laboratory's services, from crime scene examination through laboratory testing and analysis to evidence presentation in Court. It encompasses all scientific, technical and support sections of the Laboratory. Our system will be extended to include the Victoria Police Fingerprint Section, responsibility for which is being transferred to the Victoria Forensic Science Centre.

ISO Guide 25[2] defines an organisation's quality system as

'The organisational structure, responsibilities, procedures, processes and resources for implementing quality management'

Further to this, ASCLD-LAB[3] identifies the requirement for any organisation involved in forensic science to not only have a quality system in place, but also to have it appropriately documented.

The documentation of VFSC's quality system comprises six distinct manuals (refer Table 1).

VFSC-MANUAL TYPE
A. Laboratory (3 volumes + appendices)
B. Divisional Procedures
C. Section/Branch Methods
D. Section/Branch Training
E. Safety and First Aid
F. Standard Reference Works

Table 1. VFSC Quality System

3 INTERNAL QUALITY AUDITS

ISO Guide 25[2] and the ISO 9000 series of quality system standards[1] give prominence to the need for periodic internal audits. VFSC determined to adopt an on-going program of internal audits which will assure laboratory management that the quality system it has installed is, in fact, being implemented as intended.

When conducted as a scheduled series of detailed examinations of procedures and processes covered by the quality system, quality audits can be expected to reveal occasional lapses in the application of the quality system, lapses which need to be addressed through some form of corrective action in order to restore the harmony between what is happening in practice and what the quality system requires.

On occasions, internal quality audits will be conducted in response to a specific breakdown in the quality system. Such audits will be specifically directed at identifying the causes of the problem, and determining the reasons why the established quality system allowed the problem to occur. Again, such audits will identify corrective actions necessary to prevent a recurrence of the problem.

But above all, internal quality audits are a powerful means of identifying opportunities for quality improvement. The fact that procedures and methods are scrutinised in close detail may lead to the discovery of unnecessary variations, superseded requirements, inefficient practices, or simply better ways of doing things. By acting on these discoveries, laboratory management can improve the efficiency and effectiveness of the quality system (refer Table 2).

THE THREE MAJOR FUNCTIONS OF INTERNAL QUALITY AUDITS:
To MONITOR the implementation of the quality system
To IDENTIFY the cause of non-conformances and appropriate corrective actions
To DETERMINE the opportunities for improvements to the quality system

Table 2. Major Functions of Internal Quality Audits

The ISO 9000 series of quality system standards[1] clearly defines management's rôle in establishing, managing, auditing and reviewing an organisation's quality system. In particular, senior management's responsibility in auditing its own organisation's quality system has three essential components:

a) to ensure that its policy and objectives for the internal quality audit program are effectively communicated throughout the organisation;
b) to ensure that it assigns appropriate human and other resources to the internal audit program; and
c) to ensure that procedures are implemented by which feedback to management will occur in an appropriate timeframe.

4 OBJECTIVES OF VFSC'S QUALITY AUDIT PROGRAM

In defining the objectives of the Laboratory's internal quality audit program, not only was the objective of compliance monitoring fully embraced, but also the efficient and effective management of the Laboratory's resources, the promotion of staff ownership of the total quality management system, and the achievement of the Laboratory's aims, were each identified as vital components of our quality audit program.

Consequently, eight objectives were identified and adopted:

i) to assist the Laboratory in optimally satisfying its aims. In particular, the implementation of the program will assist the Laboratory to monitor its ability to fulfil the requirement to

"*ensure that the quality of scientific work carried out at the Laboratory, or under contract for the Laboratory, is in keeping with professional standards*";

ii) to provide an intra-laboratory mechanism for the timely identification of conformance or non-conformance with the Laboratory's quality system;
iii) to formalise and document auditing procedures in order to ensure that:
a) internal quality audits are efficiently and effectively planned, conducted and reviewed; and
b) where necessary, corrective actions are taken and verified within a specified time period;
iv) to identify, develop and effectively utilise the Laboratory's human resource skills base in the internal quality audit process;
v) to minimise any disruption to the areas to be audited by providing relevant staff with details of:
a) the audit program timetable for a 12 month period;
b) the names of members of the relevant audit team;
c) the issues to be audited in each audit;
d) the procedures that the audit team will follow;
vi) to satisfy ASCLD-LAB requirements for internal quality audits;
vii) to promote staff "ownership" of the quality management system by assigning key rôles to suitably trained staff members; and
viii) to provide an internal mechanism which will assist in the annual review of the Laboratory's quality system.

5 AUDIT PROGRAM MANAGER

To assist senior management fulfil its responsibilities for its internal quality audit program, the Standards[1,2] require that management allocates appropriate resources to the Program. To this end, an Audit Program Manager was appointed.

As the focal point for the coordination of the audit program, the Program Manager is responsible for a number of tasks including -

- the appropriate development and documentation of the audit procedures
- the establishment and adequate monitoring of the audit schedule
- the selection and training of appropriate quality auditors
- the development of procedures for reporting and appropriate follow up of audit results.

6 THE BEST APPROACH

The Standards[1,2] do not promote a particular approach as being the best means to audit a laboratory's quality system. Rather, their emphasis is on ensuring that management takes a pro-active rôle and evaluates each available approach in the context of their own organisational dynamics.

In determining the best approach for our organisation, four possible variations were evaluated (refer Table 3). Of these, periodic horizontal and vertical audits were determined to be most appropriate for this Laboratory, given its size and partial geographic dispersion.

Segmenting VFSC's quality system into approximately twenty "quality" issues, enabled it to be audited every 12-18 months. This audit format, with its more frequent audit schedule, was seen to be the most consistent with all of the Laboratory's objectives for its quality auditing program.

MAJOR TYPES OF QUALITY AUDITS
periodic 'whole' audit - audit of entire quality system of organisation.
periodic Divisional audit- audit of the entire quality system of one Division
periodic horizontal audit- audit of one or two quality issues (e.g. safety)
periodic vertical audit - audit of all activities associated with a particular process, (e.g. evidence handling).

Table 3. Major Quality Audit Types Evaluated at the VFSC.

One additional type of audit, an activity audit, is also possible. This is highly focussed, usually on a single activity. It is the appropriate audit for the investigation of isolated technical problems or for the routine monitoring of a vital support and/or operational activity. Significantly, a "Management Audit/Review" of the internal quality audit program has been developed and scheduled.

7 AUDIT FREQUENCY AND PRIORITY

Both ASCLD-LAB[3] and the ISO quality management standards[1,2] require that reviews of the quality system occur annually. It was decided that, in order to ensure that the annual review process is able to significantly utilise information from the internal quality audit program, our Laboratory's quality audits would be scheduled every 8 weeks and that, at least initially, a single issue would be audited each time.

The ISO 9000 quality management standards[1] assign responsibility for establishing and maintaining the quality system to senior management. Therefore, the Laboratory Director, in consultation with the Audit Program Manager, is responsible for scheduling the issues to be audited, and for varying the schedule when appropriate.

8 SCOPE OF THE AUDIT PROGRAM

The ISO 9000 standards[1] allow some flexibility as to what constitutes *'a comprehensive system of internal audits'*. We adopted a three tiered environmental scope for our audits -

level 1 - an audit of a particular Branch/Section
level 2 - an audit of more than one Branch/Section, but
not of all applicable areas
level 3 - an audit of all applicable areas of the Lab.

9 TRAINING OF INTERNAL QUALITY AUDITORS

The international quality system standards[1] require quality auditors to be trained and qualified. The VFSC determined that potential quality auditors would be selected according to the following criteria:

- communication skills
- commitment/professionalism
- analytical skills
- an appropriate cross-section of staff.

A number of training providers were evaluated according to qualitative and quantitative criteria. The National Association of Testing Authorities, Australia (NATA) was selected as the training provider. One advantage of this selection was that NATA was actively involved with ASCLD-LAB in developing the Australian forensic science accreditation program, which is now formally in place. Fourteen members of staff completed a two day training course conducted by NATA, which included an actual audit of part of the Laboratory's quality system.

The inclusion of a practical audit in the training course proved to be immensely beneficial, as the trainees were quickly able to put theory into practice, and receive timely feedback on their performance.

10 INDEPENDENCE OF AUDITORS

ISO Guide 25[2] requires that, wherever possible, quality auditors are to be independent of the activity being audited. Does this preclude internal auditors from undertaking any audits within their own organisation?

Given the nature of the work performed at the VFSC, and the need for efficient and effective use of the Laboratory's resources, it was decided that an internal quality auditor be defined as 'independent' of the activity or area to be audited if he/she does not routinely perform work in that area or activity. In our experience, an audit team of four is an appropriate size to complete the task in minimal time. During the data gathering phase, the audit team operates in pairs.

11 QUALITY AUDIT PROCEDURES

ISO Guide 25[2] clearly identifies the requirement for an organisation's Quality Manual to include 'procedures for audit and review'. Further, the ISO 9000 series[1] requires that audits and their associated actions be carried out in accordance with documented procedures.

The VFSC quality audit procedure identifies the key processes involved in the quality audit program and assigns clear responsibility for each of those processes. In particular, our procedure details the documentation to be used when planning, conducting and reporting on the audit, and the criteria by which to distinguish a major from a minor non-conformance.

The quality audit procedure also complies with the requirement that prompt action be taken to overcome the deficiencies revealed by the audit. This timely response is achieved by the development of an 'Agreed Corrective Action Plan'. This plan has an agreed timeframe for the resolution of the deficiency (non-conformance). The relevant auditor and Assistant Director are co-signatories of this plan, and are therefore bound by it.

Unlike a number of other organisations, the VFSC's corrective action procedure allows for this plan to be developed prior to the briefing of the Director by the audit team leader. This particular corrective action sequence has been chosen in an endeavour to further promote ownership of the Laboratory's quality system (Refer Diagram.1)

12 IMPLEMENTATION OF THE INTERNAL QUALITY AUDIT PROGRAM

Since the first quality audit during the training program in June, 1994, a further 7 audits have been successfully completed. Three of these audits have had an environmental scope of "level 3", with the remaining 4, "level 2".

A total of 17 staff members has been utilised. A total of 95 Agreed Corrective Action Plans has been formulated as a result of these audits, with approximately 94% having been satisfactorily completed to date. The remaining 6% are being monitored by the Program Manager and are now dependent on parties external to the original Corrective Action Plan.

As discussed earlier, one of the objectives of the audit program is to "assist in the annual review of the Laboratory's quality system". Even so, a total of 98 recommendations for improvements to VFSC's quality system have been made by the first 7 audit teams. As a response to this number of recommendations, the Director has determined that quarterly reviews of the audit program will be undertaken. The major focus of these will be to prioritise recommendations and to develop a plan for their strategic implementation.

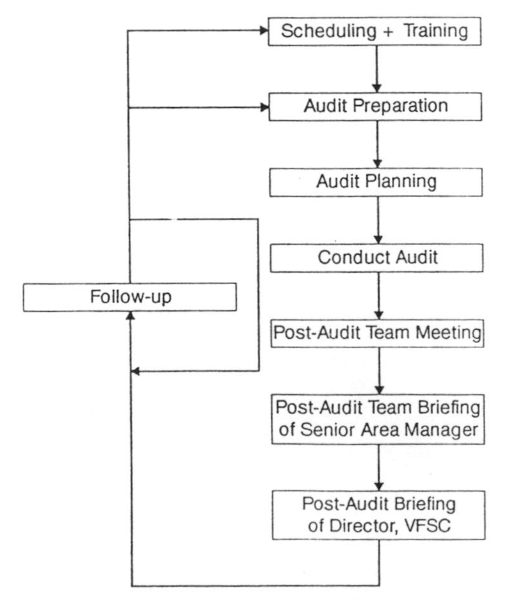

Diagram 1. The VFSC Audit Cycle

13 CONCLUSION

Although the early results of VFSC's quality auditing program are extremely promising, it is clear that two significant challenges lie ahead for VFSC management if it is to maintain the viability of the program; these are,

a) the ever increasing demands on scarce human resources which will see managers ever more challenged to determine the appropriate trade off between support and operational functions, and
b) the requirement to ensure that a sufficient pool of auditors is available given the nature and uncertainty of operational forensic commitments.

REFERENCES

1. ISO 9000 "Quality management and quality assurance standards- Guidelines for selection and use", 1987.

2. ISO Guide 25 "General requirements for the competence of calibration and testing laboratories", 1990.

3. American Society of Crime Laboratory Directors - Laboratory Accreditation Board Manual, 1994.

Sampling in the Context of TQM – with Special Emphasis on Log-normal Distributions

Erik Olsen and Frands Nielsen

DEPARTMENT OF OCCUPATIONAL HYGIENE, NATIONAL INSTITUTE OF OCCUPATIONAL HEALTH, LERSØ PARKALLÉ 105, DK-2100 COPENHAGEN, DENMARK

ABSTRACT

Monitoring the indoor or outdoor environments are examples of measurements where the variation in the object of measurement often is much larger than the variation in sampling procedure and analysis. Despite of this, most attention so far has been paid to quality control of analysis and few resources and little interest has been directed towards quality control of sampling.

For an analytical method, proper quality control can keep random errors below a few per cent for most modern analyses, if the analysis is not performed too close to the quantification limit. Regular quality control can secure a steady analytical performance and detect blunders. Participation in proficiency tests can secure against systematically erroneous analytical results. Well controlled analysis is the foundation of the total quality concept, but other elements may be of significance.

When sampling is done from a population, which has a skew and wide distribution, erroneous results may be obtained, particularly for small sample sizes. Sometimes people claim they are doing representative sampling. It is tempting to do representative sampling as costs are substantially reduced, because, if samples are representative only few samples are needed. But, representativeness of the sampling is usually not controlled.

Stratified sampling, ie. dividing the object of measurement into strata, which are known to be homogeneous concerning the measurand(s), can improve sampling efficiency and reduce costs. Stratification of a geological formation is a stratification in space only. Monitoring the pollution of a river needs stratification in both space and time. Taking samples of the air workers are breathing in the working environment needs stratification in space and time, as well as of worker mobility, the activity around the worker, and a stratification of what the worker is actually doing.

Recently a new approach, named the logbook method, for sampling at the workplace has been introduced. Using the logbook method biased data from measurement days, which are not representative for the exposure period, can be revealed. Combined bias and random errors in measurement data can be reduced, and both the short- and long-term mean exposure concentrations can established. Further processes' contributions to workers exposures are quantified. Improvements to the working environment, therefore, can be based on quantitative data.

1 INTRODUCTION

According to regulations, the workers' exposure at the workplace is measured over a full shift. Such results are denoted time weighted average concentrations (*TWACs*), because for many types of measurements, eg. measurements of organic volatiles by sampling vapours on charcoal tubes, sampling cannot be carried out over the full shift using only one tube. Consequently the shift is divided into a series of consecutive sampling periods and the *TWACs* are calculated as a time weighted sum. *TWACs* are compared with the occupational exposure limits (OELs) in force. The purpose of comparing measurement results with OELs is to secure that the air quality are in compliance with the standards.

Distributions of shift-*TWACs* are reported to be approximately log-normally distributed. Leidel et al.[1] found a geometric standard deviation (GSD) of 1.55 for American compliance measurement data of exposures to volatile organic compounds. Buringh and Lanting[2] (1991) found GSDs up to five in Dutch data, and in the most comprehensive study performed by Kromhout et al.[3], which contains about 20000 data points from five countries, the median within-worker GSD was 2.48. These findings confirm the large variability in individuals' exposure concentrations at the workplace.

Table 1 *Simulated Samplings from a Log-normal Distribution with a Arithmetic Mean of 80 and a GSD of 2.48*

	set 1	set 2	set 3	set 4	set 5
1	34	292	173	76	90
2	21	28	54	95	62
3	10	59	40	10	66
4	167	30	51	17	15
5	26	9	47	3	178
6	83	208	157	133	83
7	14	5	3	163	102
8	55	92	73	9	16
9	278	104	15	180	15
10	38	18	6	90	420
AM[1)]	73	85	62	78	104
SD[2)]	86	96	59	67	122
CV[3)]	1.19	1.13	0.95	0.86	1.17
GM[4)]	47	56	45	59	68
GSD[5)]	2.55	2.48	2.30	2.10	2.52

[1)] Arithmetic mean. [2)] Standard deviation [3)] Coefficient of variation [4)] Geometric mean [5)] Geometric standard deviation.

Table 1 illustrates the scattered data obtained, when sampling is done from a lognormal distribution with a GSD of 2.48. Calculations show that if the sampled distribution has a GSD of 2.48, then 39% of the sampled results can be expected to be below 40 (half the mean) and 11% above 160 (twice the mean). The means of 10 measurements, however, show a more moderate variation.

1.1 Number of Samples to Collect

The number of days necessary to measure, when exposures are log-normally distributed, has been given by Rappaport and Selvin[4]. To test the null hypothesis that the *TWAC* is equal to or above the OEL, against the alternative hypothesis that the *TWAC* is below, they proposed the following formula:

$$n \approx \frac{[Z_{1-\alpha} + Z_{1-\beta}]^2[\sigma_L + 0.5\ \sigma_L^4]}{[1 - \frac{TWAC}{OEL}]^2} \tag{1}$$

where Z is the value of the standard normal deviate, α is the probability of Type I error and β is the probability of Type II error, σ^2_L is the variance of the logtransformed distribution.

Table 2 shows Rappaport's[5] calculations of the sample size requirements.

Table 2 *Approximate Sample Size Requirement to the Test of the Mean Exposure for* α *= 0.05 AND (1-*β*) = 0.90*[1)]

	GSD	1.5	2.0	2.5	3.0	3.5
TWAC/OEL						
0.10		2	6	13	21	30
0.25		3	10	19	30	43
0.50		7	21	41	67	96
0.75		25	82	164	266	384

[1)] Rappaport[5].

Table 2 shows that, depending on the exposure level and the variability of exposure, the number of days required are large and measurements, therefore, are expensive.

A sufficient number of days for a confident establishment of a person's long-term mean exposure concentrations is seldom measured. Buringh and Lanting[2] had only 44 out of 420 series of measurements from Dutch industries, which consisted of 6 or more measurements on the same individual. The maximum number of measurements per individual was 13. Rappaport et al.[6] (1993) studied a comprehensive set of data consisting of 183 sets from five countries. In 55% of these datasets, the numbers of measurements per worker were below 3 and in 95% they were below 10. Only in 4.4% of the datasets, more than 15 measurements were carried out per worker.

Comparing the need as expressed in **Table 1** with the common practice accentuates that another way of measuring at the workplace is required.

1.2 The Uncertainty of the Measurement Result

The variance of a measurement result can be estimated, using the propagation of error principle:

$$\sigma^2_{TWAC} = \sigma^2_{Mea} + \sigma^2_{Obj} \tag{2}$$

where σ^2_{TWAC} is the variance of the result and σ^2_{Mea} is the variance due to the measurement procedure, ie. sampling, pump calibration, transport, storage, preparation(s), analysis, and calibration(s). σ^2_{Mea} can be established a priori and can be controlled by sampling in duplicate[7] (Taylor, 1993). σ^2_{Obj} is the variance of the object of measurement: a worker's underlying exposure distribution in the exposure period. In occupational environmental measurements the variance of the object of measurement is usually much larger than the variance due to the measurement procedure[5,8]. Because of the large variability of the object of measurement, much attention must be paid to sampling strategies, ie. where and when sampling is carried out.

2 SAMPLING STRATEGIES

2.1 Representative Samples

The most cost effective measurements would be to analyse a 'representative' sample from the population. Representative samples, however, can only be obtained from a truely homogeneous population. Mixtures of mutually miscible well stirred liquids or gases are examples of homogeneous populations, which adequately can be represented by only one sample. Despite the reported variability of exposure at the workplace, measurements in occupational hygiene are often performed with the implicit assumption that the sample is representative for the population and only one sample is taken[5,8]. The result from measuring shift-long exposure is a very accurate measure of the exposure at measurement day, but not necessarily an accurate measure of the exposures experienced by the worker on other days. It should be kept in mind that one measurement day is less than 0.5 per cent of the working time of a year.

2.2 Systematic Sampling

Rigorous execution of a sampling plan may cause that systematic relationships are found, which are artificially caused by the sampling plan. When high risk persons or processes are selected to be measured and/or sampling is carried out in periods where the exposure concentrations are high, systematically higher results are obtained[9]. For instance when measuring for control of compliance with OELs, systematic sampling is intentionally used.[10] Choosing "worst case" situations to be measured is beneficial in terms of time and money and by measuring "worst case" the occupational hygienist errs on the side of safety. This ensures that the exposure concentrations on unmeasured days are below level of concern for both measured and unmeasured workers. Results from "worst case" measurements tell nothing about the worker's mean exposure or the distribution of the worker's exposure distribution. Data resulting from systematic sampling cannot be generalized for use in the setting of standards or in epidemiology[11].

2.3 Random Sampling

Data from correctly performed random sampling is by definition biasfree, but statistically correct random sampling is very difficult to carry out. A random time period to sample, for instance, can be difficult to define. The day chosen by the hygienist to visit a facility for measuring may perhaps not be intentionally biased, but such a day is not a randomly chosen day in a statistical sense. A true random sampling plan demands that every worker has the same chance of being selected to be measured each day.

Due to financial and logistic reasons more workers are measured simultaneously during the occupational hygienist's visit in a given facility, and even workers exposures at the same workplace cannot be considered as independent.

2.4 Stratified Random Sampling

When sampling segregated (stratified) materials, such as ores of minerals in rocks, it is advantageous to use stratified sampling, ie. to divide the materials into strata from which samples are drawn by chance[12].

In the working environment, strata are more imaginary. Corn and Esmen's 'Exposure Zones'[13] and Hawkins et al.'s[14] 'Homogeneous Exposure Groups' are two examples of such strata.

In order to reduce number of measurements, Corn and Esmen[13] introduced the exposure zone principle, ie. to divide the workplace into exposure zones, in which all workers are assumed to be approximately equally exposed to the same agent(s). The assumption was based on similarity of job, environment, etc.

Recently, Hawkins et al.[14] introduced the concept of Homogeneous Exposure Group (HEG). A HEG is defined as a group of workers with identical probability of exposure to a single environmental agent. The group is homogeneous in the sense that the workers may not be equally exposed, but the distribution of exposure concentrations is the same for all members of the group. On the basis of the HEG concept, Hawkins et al. suggest that a few measurements are taken for construction of the HEG's exposure concentration distribution. From this distribution, the individuals' exposure concentration distributions are then constructed. The HEG concept is considered as the central philosophical construct for detailed workplace assessment (Ibid).

If all members in a HEG have the same exposure concentration distribution, it follows that the distribution of the group equals that of the individual members. Deduction from group to individuals, however, is logically invalid, unless workers are both exposed to the same exposure concentration distribution and that workers' exposures are independent. For instance, all or a large part of the workers in a HEG may on a busy day simultaneously experience exposure concentrations from the high end of their distributions. Then the results obtained will be a set of high results, and not the HEG's distribution.

The HEG-concept is not theoretically sound, but the concept might function reasonably well as an approximation.

Some attempts have been carried out to verify the existence of HEGs [3,5,6] Rappaport[5] found that stratification in a large series of investigations did not ensure that groups were homogeneously exposed. He defined a homogeneously exposed group (which he called a uniformly exposed group) as a group in which 95% of the individual mean exposures lie within a factor of two. Of the 31 investigations studied, 27 did not comply with the definition of homogeneity.

On the basis of these findings Rappaport[5] questioned the ability of the hygienists to assign groups appropriately solely on the basis of observation. Another interpretation, however, may be that the occupational hygienists classified correctly, but that within-group exposures simply deviated by orders of magnitudes.

Apparently a dilemma exists: A posteriori stratification as proposed by Rappaport[5], would divide the population into homogeneously exposed groups, but if workers cannot be grouped in HEGs, based on common attributes, no reduction in the number of measurements are obtained. On the other hand, a priori stratifications do not result in groups which are homogeneously exposed. Kromhout et al.[3] concluded that a priori assessment of homogeneity is not feasible and repeated measurements on the same individuals, therefore, are required. Measuring a sufficient number of days, however, is in most cases considered to be prohibitively expensive.[3,5,8]

3 AN ALTERNATIVE APPROACH

3.1 Why do Exposures Vary?

The within-worker variation found in results obtained by measuring the air concentration of a volatile chemical agent at the workplace is caused by variations in a series of variables, eg. Source strength, object worked on, tools applied, variation in the environmental conditions, the performance of the worker, the workers work pattern, ie. the fraction of time spent on different processes, production rate, and background pollution.

Table 3 *Evaporation Rates of Acetone in mg m^{-2} min^{-1}*[1)]

Air velocity (ms^{-1})	0.1	0.5	0.7
Temperature (°C)			
15	35	107	135
17	40	121	152
19	45	136	170
21	51	153	192
23	56	172	216
25	65	194	244
27	73	220	277
29	84	250	315

[1)] Calculated using the SUBTEC software package.[15]

Considering the source strength, **Table 3** shows that within the temperature range commonly met in indoor workplaces the evaporation rate of acetone varies, but not to a degree which alone can explain the reported variation in exposure levels. In the range calculated, the evaporation rate triples.

Air velocities at the workplace are commonly between 0.1 and 0.7.[16] In this range the evaporation rate almost quadruples.

The form and size of the object worked on may vary, but the influence on exposure level is limited. The tool used may influence exposure level considerably, the classical example being painting by brush or spray-gun. No occupational hygienists, however, would consider results from brush and spray painting as comparable.

Environmental conditions may change drastically from day-to-day if outside, but for modern factories with engineering controls, the indoor climate is relatively stable. One exception is when draught from open gates or windows forms turbulent eddies in the workroom air.

Production rate has been found to influence exposure level significantly, but not by order of magnitudes[17] (Esmen, 1979). The same can be said about day-to-day variations in background exposure levels.

Worker performance, *when performing the same process,* does not influence the exposure level to a degree comparable to the day-to-day variation reported.[18,19]

But in modern industry, except for assembly line workers, worker's work patterns commonly change from shift to shift, thereby changing the workers' exposure patterns.

3.2 Measuring in Process or Time Domain

When measurements are performed to establish the concentrations prevailing during periods, where several processes may be performed, the measurements are said to be carried out in time domain.[19]

Often workers perform tasks, which consist of a series of processes with distinct exposure concentration levels. When measurements are carried out in periods, where only one process is performed the measurements are said to be carried out in process domain. Stratification has been proposed, where the strata are the periods, during which a particular process is performed.[18,19] Workers' exposure concentration, measured in a series of periods during which process p is performed, is an estimate of strata p's average exposure concentration level (APC_p). The APC for each process can be assigned to each worker in the strata during times periods where he or she performed process p.

The worker's long-term mean concentration can then either be considered as the expected value in time domain (E($TWAC_j$), where j is a day in the exposure period) or it can be considered as the work pattern multiplied by the APCs.[18,19]:

$$E(TWAC_j) = TWAC_J = J^{-1} \sum_{j=1}^{J} TWAC_j \equiv \sum_{p=1}^{P} w_p \; APC_p \tag{3}$$

where J is the number of days in the exposure period, w_p is the fraction of the exposure period at which the worker truely performed process p, and APC_p is the expected value of the exposure concentration at process p in process domain. The variance of the long-term mean concentration then becomes:

$$VAR(TWAC_J) \approx \sum_{p=1}^{P} [APC_p^2 \; VAR(w_p) + w_p^2 \; VAR(APC_p)] \tag{4}$$

assuming independence between fraction of time spent at a process and the APC_p. Equation (5) shows that the variance of $TWAC_J$ is equally dependent on the variance of the time component and the concentration component. The large variance of TWAC's reported, therefore, is caused by a large variation either in the work pattern, or in the process concentrations (*APCs*), or a large variation in both.

The within-process concentration variability (VAR(APC_p)) is caused by changes in ventilation, temperature, evaporation area, background pollution, worker's performance, etc. Disregarding eg. accidental spills and working in draught or outdoors the variance of *APC* can be expected to be moderate compared with the variation in *TWACs*.[18,19] When processes must be performed within narrow limits to meet quality demands to the product, the between-worker variation in process domain can be expected to be moderate. **Table 4** shows the variation of a set of processes measured in process domain.

Table 4 *Variation of Process Concentrations*

Process	n	PC_p mg m^{-3}	SD_p mg m^{-3}	CV
Gelcoat	5	131	48	.37
Applying web	12	55	17	.31
Rolling	11	65	31	.48
At table	10	89	27	.30
Wings	12	220	96	.44
Boat	8	60	41	.68
Strip	5	189	31	.16

n is number of measurements, PC_p is the process concentration, SD_p is its standard deviation, and CV is the coefficient of variation. The coefficient of variation of measurements was equal to 6.4%, thus much smaller than the variation in the objects measured. Data were obtained on more workers at two measurement days more than nine month apart with quite different production levels.[18,19] The table shows a limited variation compared with the day-to-day variation reported (See **Table 1**). It is of note that the log-normally distributed data and the large within-worker variability reported by Kromhout et al.[3] are found in data obtained in time domain.

Time spent at a process may vary considerably from day to day. If a process is performed during 5 minutes one day and 8 hours a second day, the exposure time is approximately 100 times longer the second day. The work pattern, therefore, may vary substantially from day-to-day and it ought to be measured.

3.3 The Logbook Method

Using the logbook method, the time and the concentration components of the *TWAC*s are measured separately. Durations of processes are measured by workers keeping log about when they start and finish a process, every day in a period of several weeks, named the log period. Each process period is assigned a value for the exposure concentration during the process period (APC_p).[18,19]

The model of exposure in process domain consists of two contributions: one contribution from the background exposure and one from the process performed by the measured worker.[19] The background concentration needs not to be constant, but in case not, it must be known as a function of time. Emission source activity factors, such as developed by Franke and Wadden[20] could be used for calculating background concentrations. Activity levels can be derived from other workers' logbooks.

Working with volatiles, however, the process component of exposure is often much larger than the background component. The errors introduced in the *TWACs* by neglecting variation in the background concentration will often be minor compared with the bias that may be introduced, if sampling is carried out in the low or the high end of a wide and skewed distribution of *TWACs*.[19]

Obviously, exposure data from days, which are not representative for the exposure period, are not very useful if the bias is both significant and unknown. It has, therefore, been suggested to document that the measurement day is representative for at least a longer period of time than a single or a few days by using the logbook method.[18]

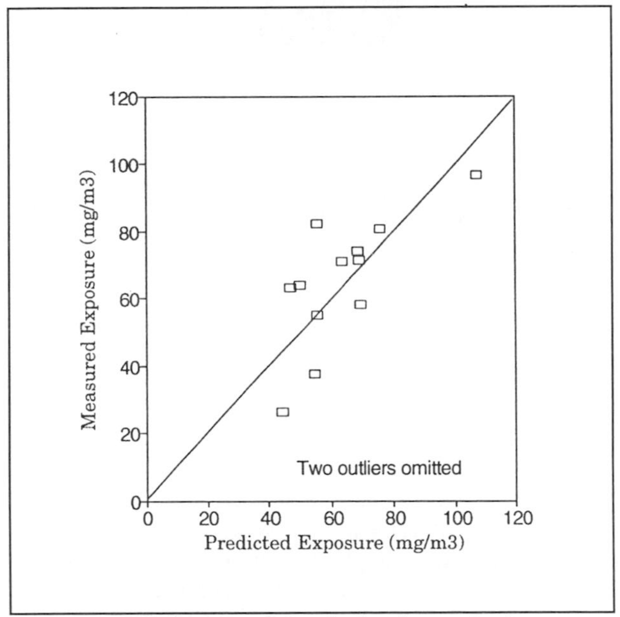

Figure 1 Validation of the logbook method.

Figure 1 shows measured exposures versus exposure predicted from logbooks kept at measurement day and *APCs* measured 23 month earlier. The two outlying results omitted could be accounted for.[19] The scatter around the identity line in **Figure 1** should be compared with the simulated results in **Table 1** and with the variability of exposure shown in **Figure 2**.

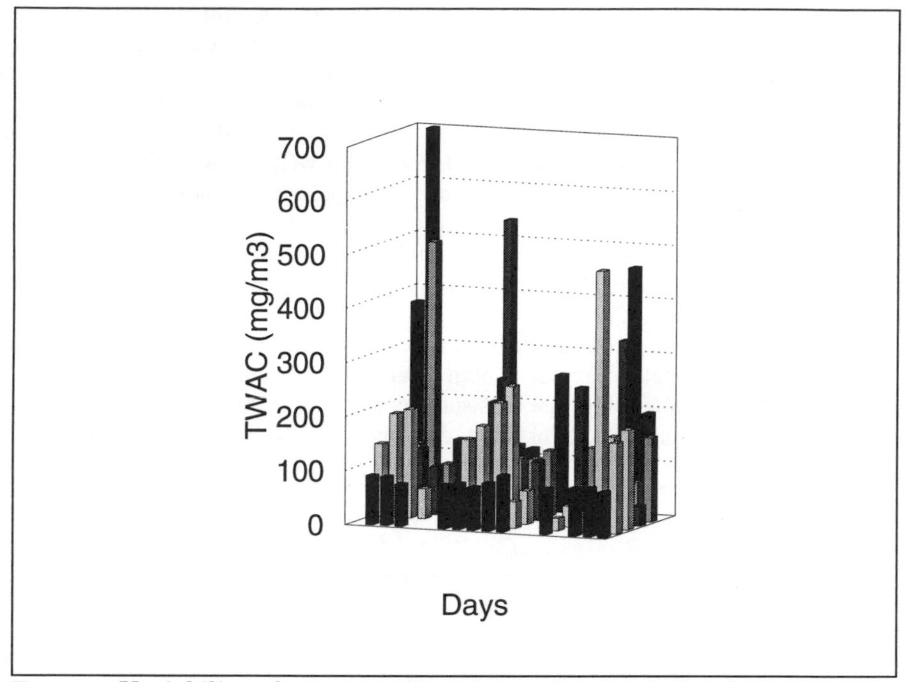

Figure 2 Variability of exposure from day-to-day for six workers.

Figure 2 shows the exposures for six workers in a three week period, measured using the logbook method.

4. CONCLUSIONS

Exposure concentrations found at the workplace are approximately log-normally distributed. The number of measurement days necessary, reliably to establish worker's exposures, therefore, is large. Such large number of measurement days is inhibited by economy and in practice many less days are measured. For small sample sizes, traditionally applied sampling strategies sometimes result in biased results, when exposure distributions are wide and skew. Focus should not only be on control of analytical performance, but also on sampling strategy. The logbook method offers a cost-effective alternative to the sampling strategies traditionally applied, because it could improve accuracy (reduces combined bias and random errors) at the expense of high precision (low random errors) when determining the long-term mean exposure to volatile chemicals for workers having shifting work patterns. Further processes' contributions to workers' exposures are quantified. Improvement to the working environment, therefore, can be based on quantitative data.

REFERENCES

1. N.A. Leidel, K.A. Busch, W.E. Crouse, 'Exposure Measurement Action Level and Occupational Environmental Variability', HEW-publication No 76-131 (NIOSH) U.S. Department of Health, Education and Welfare, Cincinnati OH, 1975.

2. E. Buringh and R. Lanting, Am. Ind. Hyg. Assoc. J. 1991, **52**, 6.

3. H. Kromhout, E. Symanski, S.M. Rappaport, Ann. Occup. Hyg., 1993, **37**, 253.

4. S.M. Rappaport and S. Selvin, Am. Ind. Hyg. Assoc. J. 1987, **48**, 374.

5. S.M. Rappaport, Ann. Occup. Hyg. 1991, **35**, 61.

6. S. M. Rappaport, Symanski, H. Kromhout, Am. Ind. Hyg. Assoc. J., 1993, **54,** 654

7. N. M. Trahey (Ed.), 'Standard Reference Materials. A Handbook for SMR Users', NBS Special Publications 260-100, Washington, 1993.

8. U. Ulfvarson, Int. Arch. Occup. Environ. Health., 1983, **52**, 285.

9. E. Olsen, B. Laursen, and P. S. Vinzents, Am. Ind. Hyg. Assoc., 1991, **52**, 204.

10. 'General Requirements for measuring Procedures. Workplace Atmospheres-General Requirement for the Performance of Procedures for the Measurement of Chemical Agents', CEN/TC 137/WG 2, 1994.

11. E. Olsen, E. Bach, N.O. Breum, J.M. Christensen, N. Fallentin, E. Holst, L. Simonsen, J. Skotte, F. Tüchsen, and P.S. Vinzents, ' Methods for the Assessment of Exposure to Chemical Agents at the Workplace', EU Commission GD V/E2, to be published.

12. American Chemical Society: 'Guidelines for Data Acquisition and Data Quality Evaluation in Environmental Chemistry', Anal. Chem. 1980, **52**, 2242.

13. M. Corn and N.A. Esmen, Am. Ind. Hyg. Assoc. J. 1979, **40**, 47.

14. N.C. Hawkins, S.K. Norwood, J.C. Rock, 'A Strategy for Occupational Exposure Assessment', American Industrial Hygiene Association, Fairfax Va, 1991.

15. E. Olsen, E. Wallström, D. Rasmussen, I. Olsen, and G. Mortensen , 'The SUBTEC-software Package Manual', Danish Working Environment Service, Copenhagen, 1992.

16. J. Gmehling, U. Weidlich, E. Lehmann and N. Fröhlich, Staub, 1989, **49**, 227 and Staub 1989, **49**, 295.

17. N. A. Esmen, Am. Ind. Hyg. Assoc. J. 1979, **40**, 58.

18. E. Olsen and B. Jensen, Appl. Occup. Eviron. Hyg. 1994, **9**, 245.

19. E. Olsen, Appl Occup Environ Hyg 1994, **9**, 712.

20. J. E. Franke and R.A. Wadden, Environ. Sci. Technol. 1987, **21**, 45.

Sampling: Are We Interested in it at All?

Pierre M. Gy

SAMPLING CONSULTANT, 14, AVENUE JEAN-DE-NOAILLES, 06400 CANNES, FRANCE

ABSTRACT :

Quality assurance or management requires analyses and tests which, with few exceptions, must be carried out on small quantities of matter : **the samples,** which have to be extracted from a batch **too bulky** to be submitted to analysis **in totality**. Hence the question : can ANY fraction of the batch be regarded as a **"representative sample"** of it ? **Certainly not !**

Sampling is not a simple handling technique : **it is a science** which, unfortunately, has not been given proper consideration until now. In this presentation of the **sampling theory** the essential concepts of **probabilistic or non-probabilistic** ; **correct or incorrect** sampling *on the one hand* ; of sufficiently or insufficiently **accurate** ; of sufficiently or insufficiently **reproducible** ; of sufficiently or insufficiently **representative** sampling *on the other,* will be introduced. Theory shows that :

* **a correct sample is structurally accurate (or unbiased)** and reliable ;
* **an incorrect sample is structurally biased and cannot be relied upon.**

We will distinguish between **(unreliable) "specimens" and (reliable) "samples".**

1 * INTRODUCTION :

Few people do realize that sampling is the foundation block on which any analytical result is built and upon which depends the **analytical reliability that is the key to Laboratory Quality Assurance (LQA)** and **Total Quality Management (TQM)**. There can be no reliable LQA or TQM without a reliable sampling, **both out-of-lab and in-lab**, and there can be no proper sampling without any reference to the sampling theory (ref [1] *in fine*).

To ignore sampling and its widely accepted rules ; to implement in your laboratory "samples" of unknown origin or liable to be biased, which we call "specimens", is tantamount to equipping a brand new car with worn-out tyres likely to blow off at any moment or to erect on shifting sands a skyscraper which will topple down at the first occasion. Sampling can be a huge source of loss or of danger.

In industry, research and trade, I can vouch that, as a consultant or expert witness, I have disclosed sampling errors responsible for losses of millions of dollars representing, in a certain case, 9 % of a company's production. Who can afford such a loss ? **CAN YOU ?**

In biology and medicine, where human life or health are at stake, I am ready to bet that a large proportion of the so-called **"analytical errors"** that are to be deplored even in the most developed countries are in fact **"undetected sampling errors"**.

Unfortunately many laboratories regard it as a simple handling technique based on the principle : **"catch whatever you can in the cheapest possible way". Quality estimation is A CHAIN and SAMPLING IS BY FAR ITS WEAKEST LINK. Experience shows that, when ignored, sampling is the most dangerous "QUALITY DESTROYER".**

YES ! We are ALL interested in sampling or, if we are not, WE SHOULD BE !

2 * IS SAMPLING A TECHNIQUE OR A SCIENCE ?

"Cheap-catch" sampling is a mere handling **technique** that consists in taking, manually or mechanically, **small increments FROM THE MOST ACCESSIBLE PART OF THE LOT** : a shovelful on top of a truck ; a scoopful on top of a drum ; a spoonful at the surface of a jar, etc. This is the manufacturers' and utilizers' favourite approach because **at first glance and in the short term** it seems to be the cheapest. The manufacturers' catalogues are full of "cheap-catch" devices that can generate the most dangerous errors. This amounts to forgetting that **sampling is a science before being a technique, even though it must implement technical solutions**. But these solutions must respect a certain number of rules that are precised by the sampling theory. Whenever these rules are disregarded ; whenever the sampling device or method is **ill designed or ill implemented, sampling becomes a dangerous "Quality Destroyer".**

3 * CONCEPTS OF PROBABILISTIC OR NON-PROBABILISTIC SAMPLING

When people speak of **"good"** or **"reliable"** samples they usually mean **"unbiased samples"** but a sampling equipment manufacturer's only means of designing a device that will extract unbiased samples is **to refer to the sampling theory** because **there is no obvious relationship between the characteristics of a sampler and an eventual bias.**

* **A batch (lot) of matter is a SET of constitutive elements** (solid fragments, liquid molecules or ions) which are assumed to remain **unaltered during sampling** .

* **The sole purpose of sampling** is to achieve a **MASS REDUCTION by SELECTION of a certain SUB-SET of elements.** We shall first distinguish between **a probabilistic and a non-probabilistic SELECTION or SAMPLING or SAMPLE** .

* A selection is said to be **probabilistic** when it gives **ALL constitutive elements** of the batch **a NON-ZERO PROBABILITY of being selected** to make up the sample.

* It is said to be **non-probabilistic** when it gives **A CERTAIN SUB-SET of constitutive elements a ZERO PROBABILITY of being selected.**

(1) Experience has repeatedly shown that **non-probabilistic sampling** generated **unacceptable sampling errors with a non-zero mean and a very large variance,**

(2) **A large number of so-called "sampling" devices** described in manufacturers' catalogues and often recommended by standards achieve a **non-probabilistic selection** that produces **structurally biased**, unreliable **"specimens"**, never reliable **"samples"**.

NON-PROBABILISTIC SAMPLING DEVICES SHOULD THEREFORE BE ELIMINATED FROM OUR PLANTS AND LABORATORIES .

4 * CONCEPTS OF CORRECT OR INCORRECT SAMPLING OR SAMPLE

A **probabilistic** sampling or the resulting sample can be **either correct or incorrect** :

* It is said to be **CORRECT** :

(a) when ALL constitutive elements of the batch are given **an EQUAL probability of being selected** to make up the sample AND
(b) when the **increments and sample** INTEGRITY is duly respected (no loss, no contamination, no alteration of any kind, etc.).

* It is said to be INCORRECT whenever one of these conditions is not fulfilled. The selection probability, though non-zero, is no longer a constant but a function of each constitutive element **physical properties** (size, density, shape, etc.). **A theory of incorrect sampling** is presented in chapter 17 of [1]. This calls for a few important observations :

(1) Correct sampling generates **three components of the total sampling error TE** :

FE : **the fundamental error,** generated by the discrete nature of ANY matter,
GE : **the grouping and segregation error** generated by the distribution heterogeneity,
IE : **the integration error** generated by the quality and flow-rate time fluctuations.

These three components have in common **a practically-zero expected value.** THEY ARE NOT RESPONSIBLE FOR ANY BIAS (chapters 13 to 21 of [1]) .

(2) Incorrect sampling generates **three ADDITIONAL components of TE** :

DE : **the increment delimitation error DE** (chapter 10 of [1]).
EE : **the increment extraction error EE** (chapter 11 of [1]).
PE : **the increment and sample preparation errors** PE (chapter 12 of [1]).

These three components of the **total sampling error** TE have a **non-zero expected value which accounts for the fact that** INCORRECT SAMPLING IS STRUCTURALLY **BIASED.** The total sampling error TE becomes a sum of six components :

TE = (FE + GE + IE) (components expressed by the **mathematical** models) +
+ (DE + EE + PE) (components tied to the **materialization** of the models)

(3) Experience has repeatedly confirmed the theoretical results : INCORRECT SAMPLING GENERATES UNACCEPTABLE SAMPLING VARIANCES .

(4) A large number of sampling devices described in manufacturers' catalogues and sometimes recommended by standards carry out an **incorrect selection** that produces **structurally biased,** unreliable **"specimens". They never produce** reliable **"samples".**

(5) Incorrect sampling devices can often (though not always) **be rendered correct** after simple modifications involving shape, size and velocity, while **non-probabilistic devices are ILL-DESIGNED and cannot be improved.**

(6) The conditions of **correct sampling** involve the **geometry and the velocity of the "cutter"** (part of the sampler which extracts the increments from the batch), as well as the general **sampler lay-out** (chapters 9 to 12 of [1]).

INCORRECT SAMPLING DEVICES SHOULD THEREFORE BE ELIMINATED FROM OUR PLANTS AND LABORATORIES .

5 * CONCEPTS OF ACCURATE OR BIASED SAMPLING OR SAMPLE

We have qualified sampling on the basis of its **conditions**. We shall do it now on the basis of its **results**. Theory shows that, for physical reasons (see [1]) **SAMPLING IS A NEVER EXACT, NEVER STRICTLY ACCURATE PROCESS.** We will now define (sufficient) **accuracy and inaccuracy** as properties of the **mean m(TE)** :

* **a sampling or the resulting sample** is said to be (sufficiently) **accurate** when the mean **m(TE)** is not larger than a certain value $\mathbf{m_o}$ regarded as an acceptable limit. Theory shows that when **sampling is correct,** the components **DE, EE, PE** of **TE** are **STRUCTURALLY zero** while the mean of **FE, GE, IE** expressed by **the mathematical models** is always small enough to be considered **negligible.** Hence the conclusion :

CORRECT SAMPLING IS STRUCTURALLY ACCURATE.

* **a sampling or the resulting sample** is said to be **inaccurate** or **biased** when the **"sampling bias" m(TE)** is larger than $\mathbf{m_o}$. The expected value of the components **FE GE, IE** (mathematical models) is **negligible. When sampling is incorrect** the components DE, EE, PE (materialization of the models) **alone** are responsible for this bias

INCORRECT SAMPLING IS STRUCTURALLY BIASED.
CORRECT SAMPLING IS THEREFORE THE ONLY WARRANT OF ACCURACY

6 * CONCEPTS OF REPRODUCIBLE SAMPLING OR SAMPLE

Reproducibility is a property of the **variance** s^2(TE).

* **a sampling or sample** is said to be **reproducible** when s^2(TE) is not larger than a value s_0^2 regarded as the **maximum acceptable.** Theory shows that when **sampling is correct,** the components DE, EE, PE are zero. The **variance** $\mathbf{s^2}$**(TE) is minimum.**

CORRECT SAMPLING MAXIMIZES THE DEGREE OF REPRODUCIBILITY

* **a sampling or sample** is said to be **non-reproducible** when s^2(TE) is larger than s_0^2. Theory shows that when **sampling is incorrect** the variances of **DE, EE, PE** are no longer zero. **The variance** $\mathbf{s^2}$**(TE) is no longer minimum.** Hence the conclusion :

ANY DEVIATION FROM CORRECT SAMPLING
REDUCES THE DEGREE OF REPRODUCIBILITY

7 * CONCEPTS OF REPRESENTATIVE SAMPLING OR SAMPLE

Many analytical procedures, many analytical standards recommend the use of **"representative samples"** but none of these has ever given an **objective, scientifically supported definition** of a **representative sample** (which I have done since 1975).

My definition of **sampling or sample** REPRESENTATIVITY is based on the properties of the **mean-square** $\mathbf{r^2}$**(TE)** which, by definition, can be written :

$$r^2(TE) = m^2(TE) + s^2(TE)$$

* **a sampling or sample** is said to be **representative** when **r^2(TE)** is not larger than a value **r_0^2** regarded as the **maximum acceptable.** Theory shows that whenever **the sampling is correct, m(TE) is practically** zero and **s²TE) is minimum** :

CORRECT SAMPLING MAXIMIZES THE DEGREE OF REPRESENTATIVITY

* **a sampling or sample** is said to be insufficiently **representative** when **r^2(TE)** is larger than **r_0^2** . Theory shows that when **sampling is incorrect, m(TE)** is no longer **zero** (the sampling is biased) **and** the variance **s^2(TE)** is no longer **minimum** :

ANY DEVIATION FROM CORRECT SAMPLING
REDUCES THE DEGREE OF REPRESENTATIVITY

For all practical purposes a sample is **representative** when it is **at the same time accurate and sufficiently reproducible** .

8 * HOW TO OBTAIN REPRESENTATIVE SAMPLES

Analytical reliability, which is a prerequisite to **TQA and TQM** demands that analysis be carried out **on REPRESENTATIVE SAMPLES** . We have just seen that, to achieve this purpose, a prerequisite is **to implement a CORRECT sampling**. When this condition is not fulfilled, the samples are **structurally biased** and **the sampling variance inflated.**

CORRECT **sampling** is therefore **a necessary condition**. To fulfil it, sampling must be designed so as to suppress the components DE, EE, PE of the total sampling error TE. I have shown in chapters 9 to 12 of [1] that these errors could be easily cancelled.

CORRECT **sampling is a necessary but not a sufficient condition.** It assures that the samples will be **unbiased** but not that they will be **reproducible enough** which representativity also demands. Thanks to the elimination of DE, EE, PE the variance of TE is reduced to the variance of **the three components associated to the mathematical models** : FE, GE, IE. Chapters 13 to 16 of [1] show how IE can be experimentally estimated. Chapters 17 to 21 of [1] show how FE and GE can be estimated and minimized.

9 * CONCLUSIONS

There are fool-proof physical tests that make it possible to tell **a genuine diamond** from **a valueless piece of strass.** There is absolutely no test, no logical piece of reasoning, that makes it possible to tell a **reliable sample** from a **valueless specimen** AS SOON AS THE LATTER IS ABSTRACTED FROM THE CONTEXT OF ITS EXTRACTION. This is probably the most puzzling paradox of sampling : **the sample or specimen does not bear in itself any trace of its degree of representativity.** No analyzer can detect, still less correct, a sampling bias or the lack of representativity of a sample. Representativity is not an intrinsic property of the sample or specimen in question but is **a property of the sampling device, method or system which has generated it.** Hence a few conclusions :

(a) To achieve LQA and TQM , the people in charge of "**Quality**" should systematically refuse to carry out analyses or tests on so-called **"samples" of unknown origin** as there is absolutely no means of assessing their degree of **accuracy, reproducibility** or **representativity**. If we except wishful thinking, the only honest way of dealing with these **"objects"**, is to regard them as **unreliable "specimens"** unless or until their origin has been carefully checked and they have been acknowledged as **reliable "samples"** by an expert eye.

(b) Considering that there is practically no fool-proof sampling equipment on the market ; considering that sampling has never been given, worldwide, the attention it deserves ; considering that the teaching of the sampling theory at university level is practically non-existent, worldwide again, etc. the corollary to (a) is that the people in charge of **"Quality"** should admit that sampling falls within their own province. They usually believe that it is somebody else's problem but they better believe me. For decades I have been looking for him : there is NO SOMEBODY ELSE. Sampling is THEIR OWN PROBLEM. **Whether they like it or not, without a reliable sampling, they will never be able to achieve Quality Management or Assurance.**

(c) To achieve these purposes, they must understand and master the sampling theory . They must organize its teaching. QUALITY SPECIALISTS SHOULD, FIRST OF ALL, BECOME SAMPLING SPECIALISTS.

The reader will find in [1] and [2] the answer to practically all questions raised by sampling in the field of LQA **and** TQM. The time has come for Analysis and Standardization to acknowledge the existence of SAMPLING as an important ANTI-QUALITY generator and that of the SAMPLING THEORY as a source of answers to most sampling problems.

I sincerely hope that, as a consequence of this Symposium THE TEACHING OF THE SAMPLING THEORY WILL BE DEVELOPED AT UNIVERSITY LEVEL for the benefit of quality-minded analysts and bodies in charge of Standardization.

10 * REFERENCES

1 - P. M. GY, Sampling of heterogeneous and dynamic material systems, Elsevier, Amsterdam, 1992 . XXX + 653 p.

2 - **F. F. Pitard,** *Pierre Gy*'s Sampling theory and sampling practice, CRC Press, Boca Raton, Florida ,1989. Two volumes 214 + 247 p.

Sampling: Are We Interested in it at All?

A. J. Poynton

KRAFT FOODS LTD, PO BOX 1673N, MELBOURNE, VICTORIA 3000, AUSTRALIA

1 Laboratory Quality Assurance and the Client

NATA (National Association of Testing Authorities) has been at the forefront of helping develop quality assurance procedures and systems within laboratories for some time. These systems and procedures go a long way towards assuring customers that the results obtained by the laboratories involved are both accurate and precise. The analytical results reported on NATA certificates have been accepted within Australia for many years and, as NATA's international reputation has grown, so too has the acceptance of these certificates overseas. This all sounds very positive, but are there any weaknesses in the process?

From the laboratory point of view, there are checks and balances at each step. When a sample arrives in the laboratory it is recorded and labelled so that it does not go astray. It is subsampled with much care, and prepared for analysis. The equipment to be used is calibrated using reference measures, then the validated test method is applied by an analyst with recently proven capability to the sample as well as further reference materials. The results are then calculated to ensure that an acceptable result is obtained for the reference material, that replicate values are within agreed tolerances, and then all calculations are checked and the results transferred to the report and the transference checked as well. At each stage the well-trained staff, without fear or favour, have followed all the procedures in their Quality System to ensure that the "true" value of the sample is obtained. The laboratory, under the guidance of its quality system is a very controlled environment, with nothing being left to chance. As far as the laboratory and NATA are concerned, the analysts have carried out their duties to the standard expected. What the client does with the result is then his problem. Or is it?

The client, on the other hand, knows that he has to convince his customer of the composition of his parcel of material. Depending on his level of expertise, he will take a sample which may be taken at random, or according to a pre-arranged plan, and assumes that it is representative of his parcel. It may be a few kilograms of iron ore which are taken from a load of several tonnes taken straight from the mine, or a few litres of crude oil from a well. He will be mindful of the cost of the analysis itself, and (in some cases) the cost of the non-returnable sample(s) sent for testing, and will naturally want to keep these to a minimum. When he receives the certificate of analysis, he promptly uses it as if it applied to all of the material. If there is a subsequent dispute over the analysis of the material, it is not only the client's reputation which is called into question but that of the laboratory and of NATA as well. And neither the laboratory nor NATA may be aware of the issue and therefore cannot defend their position.

Now it is quite understandable that the laboratory is loathe to apply the analytical result obtained on a sample to the whole shipment, especially if it has nothing to do with the sampling. Further, NATA insists that a certificate of analysis must not include statements of opinion, and opinions may abound as to whether a sampling plan was adequate or not. But the laboratory needs to be able to defend its reputation. The only avenue open is to try to become more involved in the sampling process. But sampling involves moving away from the controlled world of the laboratory into the area of probability and uncertainty. How do we know that sample of iron ore or crude oil is representative of the shipment?

2 Control of Sampling

But is sampling less controlled? In theory, setting up a sampling plan should be no different to developing an analytical method. In sampling the variables may include the number of sampling points, the quantity of material to be taken, and whether the samples should be bulked or not. As with method development, an intense sampling of typical batches is carried out to evaluate the extent of homogeneity, and then the statisticians are consulted to find the minimum number of samples required to meet the necessary level of confidence. Application of published sampling plans may also assist in achieving the level of reliability required. However, theory is rarely the real world. Sampling plans assume that the analyte is randomly distributed throughout the bulk material. This is more likely to be the case for the crude oil than for the iron ore, but by the time the laboratory had attempted to find out, the original material is on its way and a new batch with different characteristics has arrived. This hardly matches the level of control found in the laboratory under quality systems, but it is important to tackle it properly. A poor sample could result in a load of quality ore being sent for expensive secondary processing, rather than going straight to the smelter.

An even more problematic situation is checking batches of peanuts for aflatoxin. Aflatoxin is a liver carcinogen produced by a mould which can grow on peanuts, and consequently of concern to public health. A maximum limit of 15 μg/kg aflatoxin has been legislated in Australia; however single nuts have been analysed at levels well above 1,000 μg/kg. The sampling plan required to detect such contaminated nuts in a load requires a lot of samples. The excellent work which now allows detection at 0.1μg/kg or less may be wasted on a handful of contaminated nuts, but it is useful in allowing many samples to be combined at the detection stage of the analysis.

3 Sampling vs Analysis

This leads us to look at the relationship between sampling and analysis, and the impact each has on the our ability to find the true value of the bulk material. Since the variance of the whole is the sum of the component variances, we can write

$$s^2_{value} = s^2_{sampling} + s^2_{analysis}$$

where s^2_{value} is the reliability with which we can determine the true value of the bulk lot
$s^2_{sampling}$ indicates how representative the sample is of the lot
and $s^2_{analysis}$ indicates the precision and accuracy of the analysis

It follows then that if the standard deviation of the sampling is more than three times that of the analysis, then the analytical error will contribute less than 10% to the overall variance. In such a case, finding ways to increase the sample size or number of samples will have more benefit that trying to improve the analytical accuracy. Indeed a less precise but more readily accessible method may allow more samples to be analysed for the same effort with an overall improvement in confidence in the final result. The principles of TQM exhort us to get closer to our clients for our mutual benefit. The arguments above show that there is a very sound basis for this.

4 Sampling and the Internal Client

But it is not only in the external environment that we have to understand the role sampling plays in the interchange between a laboratory and its customers. In a manufacturing facility, the Quality Control laboratory will certainly be a lot more involved in the sampling process. Often, the operator of the facility will not just be interested in the analytical result to assess quality, but will be also use it in his accounting system to manage costs and losses. The adequacy of sampling will therefore be much more obvious. But in manufacturing other pressures and influences can also make representative sampling difficult.

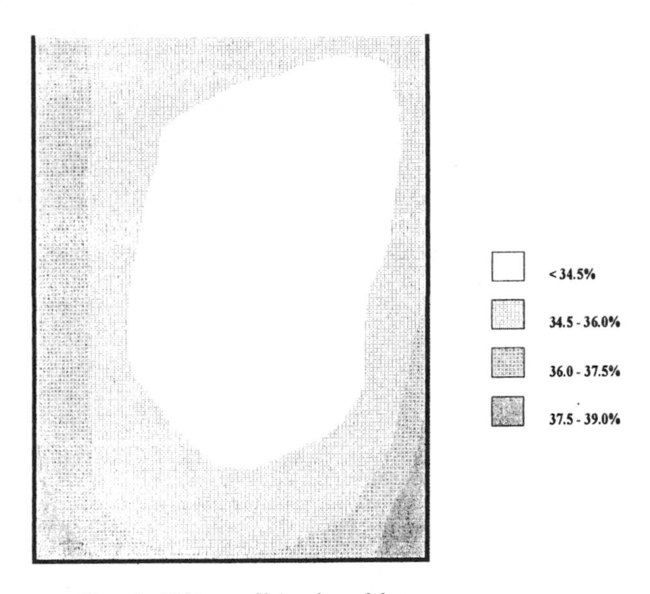

Figure 1. ***Moisture profile in a drum of cheese.***

Take cheese manufacture, for instance. The composition of milk changes due to many factors such as season, weather, time of milking, etc. These changes are not always easily detected in the milk itself, but may become evident in the cheesemaking process and/or the final cheese. It is therefore important to analyse the cheese as soon as it is made in order to adjust the process for subsequent batches. Herein lies a problem. The final stages of cheesemaking involve salting the chips of curd, draining off some "white" whey, then pressing the curd to release even more whey. The proximate analysis of the curd should now represent its final composition, and of course that is the information the cheesemaker needs as soon as possible. It is important to only take a small sample (holes left in the cheese at this stage provide a haven for mould growth), but the salt and moisture are not evenly distributed through the chips yet. Further, the pressing and draining well may have allowed the release of some of the unwanted moisture, but the distribution of the rest is far from uniform. Figure 1 shows the moisture profile in a drum of cheese immediately after draining, and we have found up to 6% difference within a single drum of cheese. The question is where do you take the sample to ensure it is representative? Is the moisture profile the same for each drum? Precise analysis at this stage of the process is a waste. After the cheese has been allowed to develop for 10 days or so, the moisture and salt are much better distributed and a more representative sample may be taken. Standard sampling plans and analytical methods can then show up the need for alterations to the cheesemaking procedure, but a lot of less than optimum cheese can be made in the meantime.

5 The Challenge Ahead

As we can see, representative sampling can be difficult. Statistical sampling plans are not particularly easy to use effectively as they depend on the distribution of the analyte being uniform. Chemical analysis is destructive and expensive, with the consequence that, due to economic pressures, the plans are rarely adhered to anyway. Many clients do not understand statistics or probability - they just want to know the facts. It is no wonder that laboratories have tried to ignore sampling. But if we continue to do so, our analytical and internal quality assurance efforts will be wasted. Our relevance to society and our reward from it are dependent on taking a leadership role in this area.

Our clients want to know what is in their bulk materials, and as we have seen, that requires both appropriate sampling and analysis. The challenge for the laboratory is to develop the skills and expertise to be able work with its clients to achieve this in a manner which is cost-effective for both parties. Even more of a challenge is to fit sampling of bulk materials into laboratory quality systems at a level of reliability comparable to that of our analytical procedures. It will require the involvement of third-party accreditation bodies, such as NATA, to assist and guide this process as they in turn will have to tackle how they are going to assess whether an appropriate level of capability and control has been achieved.

Whether they are interested or not, laboratories need to be more than just interested in sampling; they need to be actively involved. The laboratory's ability to understand and work with its clients on managing sampling issues will be a key to its long-term viability.

Sampling: Is it a Weak Link in Total Quality Management?

A. M. Henderson, Formerly Senior Lecturer, RMIT University

NATA ASSESSOR FOR CHEMICAL TESTING LABORATORIES, PO BOX 32, POINT LONSDALE, VICTORIA 3225, AUSTRALIA

1. INTRODUCTION.

Clearly the quality of any product primarily depends on the quality of the raw materials used for its production. In materials testing and quantitative analysis these comprise the certified reference materials used, analytical reagents, and the samples submitted for analysis. A laboratory has direct control of these except for the samples suppled by clients for testing or analysis. If a sampling error is present it is transposed to the final result, which in spite of NATA accreditation, has little meaning. How many NATA registered laboratories have ever assessed the sampling procedures used by their clients or those used in their own organisation?

Fortunately many materials which are tested or analysed are gaseous or a single liquid phase, and present no sampling problem. However, large solid masses which require drilling, multiphase aggregates, and particulate solids, i.e. comprising multiple particles of different sizes, are more difficult to sample because particles of different densities and sizes tend to segregate during handling . With such materials random sampling, to be defined shortly, is essential to avoid bias.

Cost considerations often determine the precision achievable in primary sampling, e.g. diamond drilling of a mineral deposit. This is taken into account by subsequent statistical analysis of the data. However, it is during the later stages of resampling, used to reduce the bulk of a primary sample, that undetected problems can occur. Bias can be introduced, and the random sampling error, which is always present, can be inadvertantly increased. The magnitudes of both types of error are highly dependent on the sampling equipment and procedures used.

During resampling all solid samples are particulate; i.e. they consist of a large number of solid particles of different sizes ranging from fine dust to the coarsest particle present. Apart from samples used for measuring particle size dependent properties, each resampling stage comprises particle size reduction followed by removal of a smaller sample. This sequence is repeated until the final sample for testing or analysis is obtained, which may be only a fraction of a gram. At each stage the maximum particle size present determines the minimum mass which can be extracted as a representative sample in the following stage. This will be quantified shortly.

2. THE NATURE OF SAMPLING ERROR.

Error in sampling is defined as the true value of a property in the sampled lot subtracted from its value in the sample. It may be positive or negative and of course is unknown. It is the sum of two components, bias and random error. Bias has an assignable cause, whether known or unkown, and for a specific sampling procedure is always either positive or negative. Random error results from chance occurrences and can be positive or negative, with an average which approaches zero as more replicate samples are tested. It depends on the size of the largest particles in the sampled lot and on the mass of the sample, not , as many people believe, on the proportion of the total material taken as the sample. The smaller the largest particle and the larger the mass of the sample the lower will be the random error. This dependence is expressed by the so-called safety rule developed by Pierre Gy[1], which is

$$Ms >= 125000d^3$$

Where Ms is the minimum mass of sample required and d is the diameter of the largest particle in the material to be sampled.

If the largest particle is 2 mm, i.e. 0.2 cm, the minimum sample size to be representative is

$$125000\ (0.2)^3 = 1000 \text{ gram}$$

Note that Standards Aust. AS 1141.3[2] specifies a minimum sample of 2000 gram for concrete aggregate of 2mm or less in size, probably because properties which are surface area dependent are so important for this material.

Random sampling is required if segregation is present. For discrete items it is simple in concept. All items present must have the same probability being included in the sample.

Random sampling of particulate materials is more complex and is defined as follows. The lot to be sampled is envisaged as comprising a large number of equal volume elements with their shortest dimension at least 3 times the diameter of the largest particle present. A number of these elements is then removed such that every element present has the same probability of being selected. The selected elements are combined to form the sample. In practice, it is not possible to delineate exactly equal volume elements because each one must contain a whole number of particles. Particles on the boundaries between adjacent volume elements pass at random into either the sample or reject, thereby marginally increasing the random error.

3. SAMPLING EQUIPMENT.

Much of the sampling equipment currently available, and widely accepted for use in industry introduces bias, which may or may not seriously affect the final measurement. The following two simple general principles are recommended for selecting sampling equipment . Avoid equipment for which the physical location of any particle in the feed determines whether it will pass either into the sample or the reject. Avoid stream sampling equipment for which the collector device does not remove the material from all points across the stream for the same time interval. Such equipment includes table and spear samplers for bulk materials, and for streams, flap samplers, hose samplers, poppet valves, and rotary samplers other than those with collector openings shaped as sectors of circles centred on the axis of rotation. All of these types introduce bias when sampling segregated material.

The most desirable type equipment for resampling is that which converts the lot to be sampled into a continuous flowing stream, or a stationary elongated strip of uniform cross section. Approximately equal volume elements can then be extracted from any part of the stream or strip as desired. To remove equal volume increments from a flowing stream it must have a constant volume flowrate, and the sample collecting device which enters the stream must extract material from every point in its cross section for the same time interval. The motor driven rotary sampler, illustrated schematically in Figure 1, satisifies this condition.

A series of collecting chutes shaped as adjacent sectors of a cylinder rotates at constant speed below a bin from which the material is fed as a stream at constant volume flowrate. Each sector passes through the stream and samples every point in the stream for

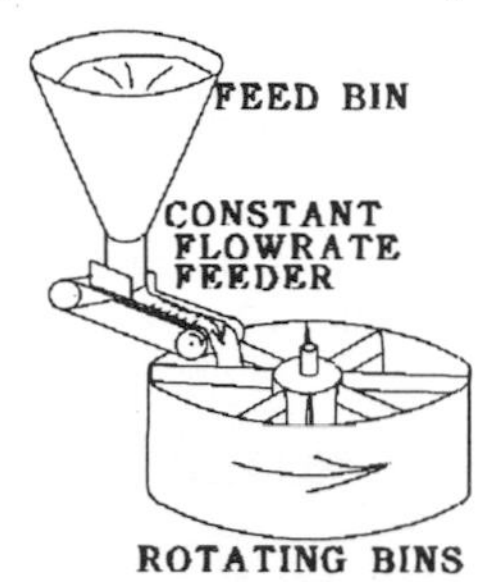

Figure 1. *Laboratory Rotary Sampler*

the same time interval to yield a small constant volume increment for each sample. Many increments are combined to form each sample of which usually eight or sixteen are produced.

The commonly used methods of coning and quartering, and the Jones riffles, the latter illustrated in Figure 2, represent a different approach.

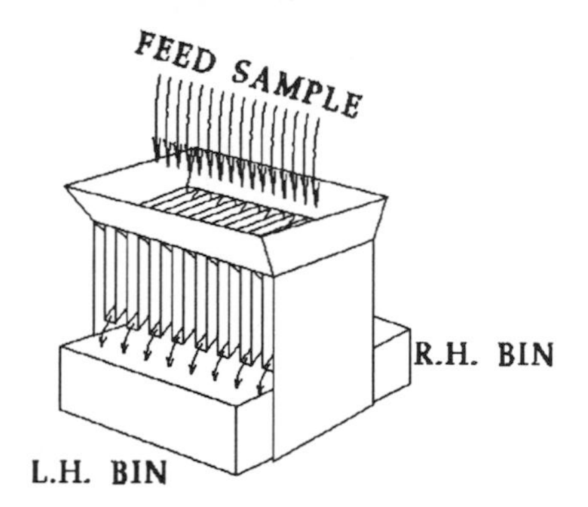

Figure 2. *Jones Sampling Riffles*

Here the aim at each sampling stage is to divide the sampled lot into two equal parts. Repeated resampling is usually required to produce the final desired sample size.

This scheme has the advantage that less total crushing and grinding are needed, but at the cost of increased random sampling error, because the total sample variance is the sum of those produced by each stage. If desired this increase can be avoided by grinding the feed to each resampling stage somewhat finer.

Riffles have a maximum of about 20 slots. Material fed to it is split at most, into 20 separate volume increments which are combined into two samples each of 10 increments. This is far less than the number produced by the rotary sampler and the random error is correspondingly higher. This error, however can be reduced by mixing of the material between each sampling stage; the more thorough the mixing the lower is the error. As will be demonstrated shortly this interstage mixing is also required to avoid biased sampling when using riffles.

4. SOURCE OF BIAS IN SAMPLING WITH RIFFLES.

To avoid bias the riffles must be used in the manner intended by their designer. Each set is provided with a feeding tray on which the sample is intended to be mixed and spread before it is fed to the riffles. However, in how many laboratories is the feeding tray used? It makes an already labour intensive operation even more so! The more common practice is to feed the riffles by pouring directly from the collecting bin which contains the sample from the last splitting stage. This practice causes significant bias in the particle size distribution sample obtained in the next stage, and a corresponding bias in all properties which are particle size dependent, as will now be demonstrated.

Figure 3 shows a longitudinal cross section of one the collecting bins in place while the riffles are being fed.

Note that only alternate slots discharge into the bin. Consequently the discharged material starts to form mounds centred under the slots with valleys between them. As the material accumulates it flows down the sides of each mound. The coarser and denser

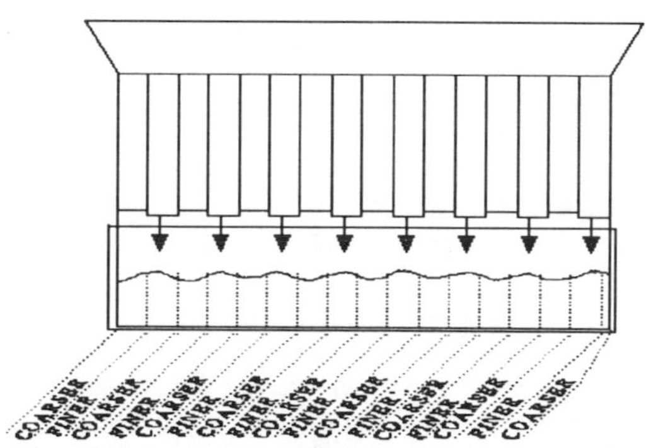

Figure 3. *Section through a Receiving Bin of Jones Riffles*

particles with higher momentum tend to bounce sideways and become concentrated around the base of the mounds, while finer lighter particles tend to heap directly below the discharging slots. This segregation is apparent in Figure 4.

Figure 4. *Particle Size Segregation in Riffled Sample (from a unit with 16 slots)*

This process continues for most of the time that the material is discharging into the bin, resulting in alternating vertical bands of coarser and finer particles along the length of the bin. When the bin is used to pour the sample for the next split the coarser bands are aligned with all the slots which discharge into one side of the riffles and the finer bands with the slots which discharge into the other side. The amount of segregation is less, see Figure 5, but there is clearly a larger proportion of coarse material in the right hand bin which demonstrates the particle size bias resulting from this method of sampling.

Figure 5. *Particle Size Bias in Samples after Second Pass through Riffles*

5 STREAM SAMPLING WITH JONES RIFFLES..

If the material to be sampled is poured into the riffles from one corner of the bin as a uniform stream which is passed back and forth over the slots, the action simulates that of the previously considered rotary sampler. The sampled material is divided into 16 small volume increments with every pass, each in turn being delivered to alternate sides of the riffles. If 50 such passes are made, the number increments is 400 per sample compared with 8 for the standard riffling method. Bias is avoided because the regions of segregated particles from the previous pass through the riffles are now each divided into many small portions which are distributed in approximately equal quantities between the two resulting samples. The random error should also be reduced by the increased number of increments being combined to form each sample.

For a period of eight years these aspects of riffling were investigated by the second year students in the Metallurgical Engineering course at RMIT. Without exception their experiments demonstrated that the observed bias significantly affected the results, and that stream riffling produced signficantly less random error than the standard method of feeding the riffles. Briefly the experiment was as follows. A 6.5 Kg sample of sand was reduced to produce 3 sets of 10 replicate 100 gram samples which were sieved at 500 μm. Six passes were made through riffles using each of the following procedures; a) the left hand bin was always accepted to give the finer sample; b) left and right hand bins were alternately accepted to give the coarser sample; c) stream riffling was used. Statistical significance testing on the results for 1968 which were typical, yielded a t value of 2.97 for the difference in mean % coarser than 500 μm for the coarser and finer samples. The critical value for 0.0025 probability is 2.947. The f values comparing procedures a with c and b with c were 80.5 and 33 respectively, The critical f value for 0.01 probability is 3.6.

6. ASSESSMENT OF A SAMPLING PROCEDURE.

How can a sampling procedure be assessed? What will be the precision of the final result? Clearly the absolute error in an individual test result is unknown, otherwise every result could be corrected. We cannot even state with certainty that the error will be within specific limits e.g. + /- 0.3%. The best that can be achieved is to state the probability that the error will be within specified limits, e.g. in 95% of samples on average the error will within the limits of say +/- 0.3%. Even this degree of certainty is impractical in most cases because it involves testing or analysing at least 30 replicate samples of the same lot. However, by allowing a further degree of uncertainty the number of replicate samples which must be tested is reduced to about 5 to 10 . The precision of the results is then expressed as follows; in 99% of the total results there is a 95% probability that the error will be within +/- 0.3%. Note that this is not the same situation as 95% of of the total results having a probability of 99% of being within +/- 0.3%. In specifying tolerances both probabilities must be quoted. The tolerance limits are estimated by replicate sampling of the same original material after which each sample is tested or analysed by the same procedure, which must be carefully controlled because it makes its own contribution to the total random error of the final result. The tolerance limits are then expressed as

$$\bar{X} \text{ +/- } kS$$

where $\bar{X}$ is the mean of the samples and S is their standard deviation.

The values of k are obtained directly from tolerance distribution tables[3,4]. The value of k depends on n, the number of replicate samples used, on γ, the confidence level, and $1 - \alpha$, the proportion of total values falling within the tolerance limits.

The simplest way to illustrate the method is by working an example, viz. the stream riffling done in the RMIT experiment.

Here let's use the values $\gamma = 95\%$ and $1 - \alpha = 99\%$

S, the sample standard deviation is $\sqrt{0.04} = 0.20\%$

The value of k from the tables for $n = 10$ replicate samples is 4.443

The appropriate tolerance limits for the % coarser than 500 µm are then

$$67.0 \text{ +/- } 4.443 \times 0.20 = 67.0 \text{ +/- } 0.9\%$$

This value may be too high for some purposes. It was caused by the small sample of 100 gram used for the sieving due to the time constraints on laboratory classes.

Gy's safety rule calls for a minimum sample weight of 1000 gram for the material which had a maximum particle size of 2 mm. The expected tolerance limits for this sample size are about +/- 0.3%, but this would need to be confirmed by experiment.

7. WHERE TO NOW?

One of the conditions for issuing NATA endorsed reports is that all standard samples used for preparation of standard solutions and calibration of equipment must be traceable back to certified reference materials, if they are available. The methods of standardising, and calibrating must be fully documented and the documentation controlled. It seems inconsistent that no similar requirement is placed on samples which are submitted by clients for analysis or testing. A sampling error is no different in its effect on the final result from an error in standardising materials or calibrating equipment.

Should a NATA endorsed certificate for sampling done by any client be a prerequisite to issuing NATA endorsed report?

This would require that each sample submitted for testing be traceable back to its primary source through all stages of resampling. Procedures and sample weights would be fully documented, and a reliable estimate of the the sampling variance for samples from each source would be provided. This would allow tolerance levels to be included in Nata endorsed reports, a common practice in engineering and science. The cost of running the once-off tests or analyses on 5 to 10 replicate samples for each sample source, would be a marginal addition to the already incurred cost of complying with the other conditions of NATA accreditation. The organisations submitting the samples would have valuable feedback for assessing the adequacy of their sampling procedures. A check on sample variance might well be run periodically on samples showing a long term trend of increasing variability. All this would surely help achieve total quality management in area which apparently has received less attention than it warrants.

Here is a challenge for us all. Will we respond, or isn't anyone interested?

References.

1. P.M. Gy 'Sampling of Particulate Materials , Theory & Practice' Elsevier, Amsterdam, Oxford, New York, 1979.

2. Standards Australia AS1141.3, 1986, 'Sampling and Testing of Aggregates.'

3. R.E. Walpole & R.H. Myers 'Probability & Statistics for Engineers & Scientists 2nd Ed., Collier MacMillan, London & New York, 1977.

4 C. Eisenhart, M.W. Hastay, & W.A Wallis, 'Techniques of Statistical Analysis', McGraw Hill, New York, 1947.

Proficiency Testing: A Laboratory Quality Improvement Tool

C. M. van Wyck

ACIRL, JUNCTION STREET, TELARAH, NSW 2320, AUSTRALIA

Preamble

According to 1S0 guide 43 (Development and Operation of Laboratory *Proficiency Testing*), *Proficiency Testing* "is the use of results generated in interlaboratory tests comparisons for the purpose of assessing the technical competence of participating testing laboratories".

Proficiency Testing may also be used to determine the precision values i.e. repeatability (r) and reproducability (R) for a particular test or for a number of tests.

Acceditation bodies such as N.A.T.A. use the results obtained through *Proficiency Testing* in their assessment of technical competence. Satisfactory performance is a condition for continued registration.

The greatest benefit to be devised from participating in *Proficiency Testing* programs is to make use of the program and the results derived therefrom as the foundation for laboratory quality improvement.

INTRODUCTION

A.C.I.R.L. (Maitland), primarily a coal testing facility, participates in the following *Proficiency Testing* programs:

i)	N.A.T.A.	37 Participating Laboratories	3/Annum
ii)	Coal Producer (Customer)	21 Participating Laboratories	4/Annum
iii)	H.V.C.L.A.G.	14 Participating Laboratories	4/Annum

Additional to the above programs, A.C.I.R.L (Maitland) runs a *Proficiency Testing* program with some twentyone (21) participating laboratories, made up from seven (7) A.C.I.R.L. laboratories, four (4) Australian coal laboratories and twelve (12) overseas coal laboratories.

WHY PARTICIPATE ?

Participation in *Proficiency Testing* is twofold. Firstly compulsion and secondly as a quality improvement tool.

A.C.I.R.L. (Maitland) participates in the N.A.T.A. program for a combination of the above two reasons. As a supplier of samples to N.A.T.A. for use in their *Proficiency Testing* programs, we use the homogeneity data obtained as a Quality Measure in respect of our sample preparation and sample division procedures and we participate in the subsequent test work because of accreditation requirements.

Participation in the Coal Producer (Customer) *Proficiency Testing* program is for the first reason i.e. compulsion. As a supplier of laboratory services we are required to participate in the Customers program in order to maintain 'Approved Supplier' status according to our Customers Total Quality Management requirements.

Participation in the A.C.I.R.L. and H.V.C.L.A.G. programs is for the second reason i.e. as a quality improvement tool.

HOW PARTICIPATION OCCURS

This heading might better read as 'How Participation Should Occur'.

The philosophy underpinning *Proficiency Testing* programs is that the samples are treated exactly as are routine samples submitted to the laboratory. When compulsion is the reason for participation, there is great pressure on laboratory management to deliver the 'correct' numbers. Thirty plus years experience as a tester, manager and laboratory assessor have taught me that the pressure is too great for many laboratory managers and *Proficiency Testing* program samples are subjected to 'special' treatment.

Only when the reason for participation is because of a genuine desire to use the program as a Quality Improvement Tool and the philosophy of treating the samples as routine, can greater value be attributed to the results obtained.

PROGRAM DESIGN

The A.C.I.R.L. (Maitland) program is designed to cater for A.C.I.R.L. requirements. The parameters participating laboratories are asked to perform is the range of tests which are most commonly performed in our company's laboratories. There is no compulsion on participating laboratories to do all the tests on all of the samples submitted to them.

Table 1. Shows the range of tests in the program and is a fair representation of the average participation rate.

Samples are chosen to provide a suitable range of values for the varying parameters. Naturally, with the material used in our program i.e. coal, it is not always possible to have suitable properties for all of the parameters: e.g. Coking and Caking properties are generally at the lower end of the scale when we wish to test for say chlorine. For coals displaying coking and caking properties, we generally have coals with chlorine values in the range of

0 to 0.01%. We may desire to learn if we are proficient at testing for higher chlorine values than this small and limited range. It may therefore be necessary to sacrifice testing for the coking and caking parameters (such as Crucible Swelling Number, Gieseler Plastometer and A-A Dilatometer) for a particular round of the program.

Because of its nature, coal is a very non-homogeneous substance, displaying a wide variation in properties in relation to particle size. Therefore, sample preparation and division is a major component to consider when designing a *Proficiency Testing* program.

The A.C.I.R.L. program provides samples at minus 212μm and at minus 4.0mm. Homogeneity testing of these materials is a valuable Quality Tool.

PRESENTATION OF RESULTS AND THEIR USEFULNESS

The format adopted for presentation of *Proficiency Testing* results is a matter of choice for the program manager. A.C.I.R.L. adopts the K.I.S (Keep It Simple) principle i.e. tabulating the results from participating laboratories together with a mean result and standard deviation (see Table 1). It is up to the managers of the participating laboratories on how best to make use of the available data.

N.A.T.A. *Proficiency Testing* program reports are much more elaborate, identifying stragglers, outliers, methods etc and in addition to tabulating results also display the results by histogram (see Figure 1) and on Youden Diagrams.

The laboratory manager armed with the data devised from *Proficiency Testing* participation is now in the position to quantify the proficiency of his laboratory in respect of test method procedures and if he so desires the operators.

A further benefit to be devised from operating ones own *Proficiency Testing* program is to analyse the data, applying the values for properties of interest to reserve material from the sample preparation stage and using this material for 'System Process Control' as a Quality Control measure: (see figure 2)

CONCLUSION

Proficiency Testing in the laboratory is a very valuable Quality Improvement and Measurement Tool.

This value is maximised by controlling the program because of :

- i) Choice of test parameters
- ii) Choice of test material
- iii) Sample preparation and division homogeneity data
- iv) Provision of reserve material for SPC charting.

Reference: ISO GUIDE 43

Development and Operation of Laboratory Proficiency Testing

Table 1

ACIRL RR QUALITY PROGRAM

SAMPLE: November 1994

Laboratory / Analysis																							Mean	Std. Dev.
Moisture (as analysed)	%	2.5	2.7	2.1	2.1	2.3	2.8	2.8	2.9	3.1	3.0	2.9	2.8	2.8	2.8	2.8	3.2	2.8	3.1	2.5	2.8	2.2	2.7	0.311
Ash (d.b.)	%	9.4	9.4	9.5	9.4	9.4	9.5	9.5	9.5	9.5	9.4	9.6	9.5	9.4	9.5	9.6	9.5	9.5	9.4	9.4	9.5	9.4	9.5	0.054
V.M. (d.b.)	%	34.0	34.2	34.6	34.4	34.3	34.7		33.9	34.4	34.2	34.3	34.4	33.9	33.7	34.5	33.6		33.6	34.1	34.2	33.6	34.1	0.334
F.C. (d.b)	%	56.6	56.4	55.9	56.2	56.3	55.8		55.6	56.1	56.4	56.1	56.1	55.7	55.8	55.9	56.9		57.0	55.5	56.3	57.0	56.4	0.353
Total Sulfur (d.b.)	%	0.45	0.50	0.48	0.56	0.46			0.47	0.43	0.49	0.48	0.48	0.41	0.47	0.48	0.40	0.45	0.45	0.49	0.48	0.51	0.47	0.035
Chlorine (d.b.)	%	0.04			0.20					0.06	0.04						0.04						0.08	0.062
Phosphorus (d.b.)	%	0.018								0.019													0.019	0.001
Carbonate Carbon (d.b.)	%	0.03	0.02							0.03													0.03	0.005
Specific Energy (gross)(d.b.)	MJ/Kg	30.92	31.05	31.22	30.81	31.23	31.33		31.13	31.05	31.41	31.30	31.15	31.19	31.42	30.85	31.12		31.40	30.98	31.16	31.16	31.15	0.174
Specific Energy (gross)(d.a.f.)	MJ/Kg	34.13	34.27	34.50	34.01	34.47	34.62		34.40	34.31	34.67	34.62	34.42	34.43	34.72	34.14	34.39		34.65	34.19	34.43	34.39	34.41	0.194
Relative Density	-	1.36	1.35	1.35	1.36		1.35			1.34	1.34	1.36			1.35	1.35				1.33			1.35	0.009
Ultimate - Carbon (d.a.f)	%		83.5							85.1	83.2				82.6	82.6				82.9			83.3	0.859
- Hydrogen (d.a.f)	%		5.52							5.50	6.31				6.30	5.74				6.80			6.03	0.478
- Nitrogen (d.a.f.)	%	2.31	2.08							2.44	2.14				1.81	1.97				1.94			2.10	0.203
- Total Sulfur (d.a.f.)	%	0.50	0.55	0.53	0.62	0.51			0.52	0.48	0.54	0.53	0.53	0.45	0.52	0.53	0.44	0.50	0.50	0.54	0.53	0.56	0.52	0.039
- Oxygen (d.a.f.)	%		8.35							6.48	7.81				8.77	9.16				7.82			8.07	0.856
H.G.I.	-	47			53					47	51		56	50	56	49	48		49				51	3.200
Ash Fusion (red) - D.T.	degC	1340			1310					1340	1360					1460	1400		1440				1379	51.942
- S.T.	degC	1490			1480					1490	1470					1550	1550		1510				1506	30.170
- H.T.	degC	1530			1500					1510	1510					1590	1570		1520				1533	31.493
- F.T.	degC	1570			1555					>1550	1570					>1600	1590		1550				1584	31.703
Crucible Swelling Number	-	5.0	5.5					3.5	5.0	6.0	2.5				4.0	3.0				4.0			4.3	1.105
Giesler - I.ST.	degC		400	395					400	400													399	2.165
- M.F.T.	degC		450	440					435	435													440	6.124
- S.T.	degC		470	460					460	460													463	4.330
Range	deg		70	65					60	60													64	4.146
M.F.	dd/m		70	100					60	80													78	14.790
M.F. (log10)	-		1.85	2.00					1.78	1.90													1.88	0.080
Dilatometer - T1	degC		405							405													405	0.000
- T2	degC		435							440													438	2.500
- T3	degC		470							450													460	10.000
- c	%		36							30													33	3.000
- d	%		-3							6													2	4.500

Figure 1

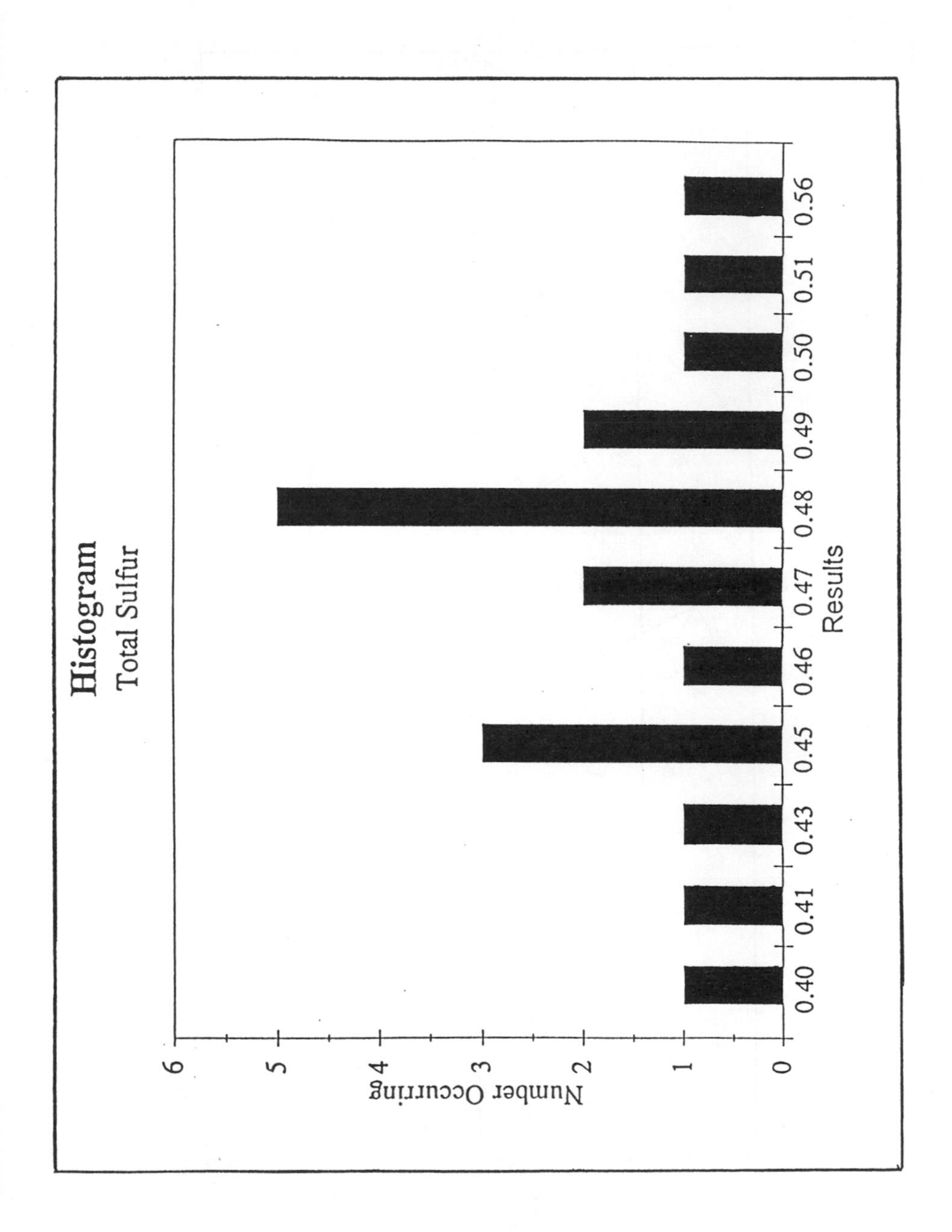
Histogram
Total Sulfur
Number Occurring
0
1
2
3
4
5
6
0.40
0.41
0.43
0.45
0.46
0.47
0.48
0.49
0.50
0.51
0.56
Results

Figure 2

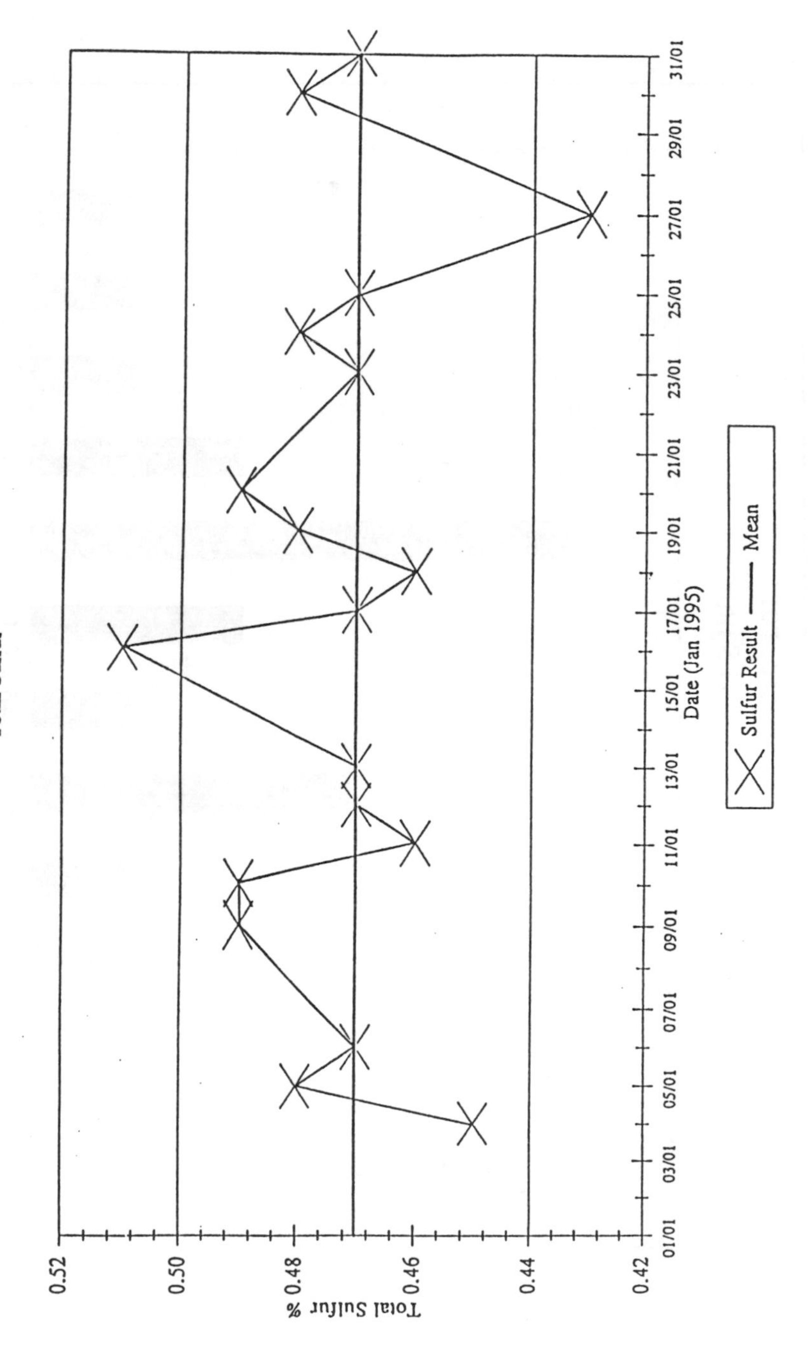
SPC Chart
Total Sulfur
Total Sulfur %
0.52
0.50
0.48
0.46
0.44
0.42
01/01
03/01
05/01
07/01
09/01
11/01
13/01
15/01
17/01
19/01
21/01
23/01
25/01
27/01
29/01
31/01
Date (Jan 1995)
Sulfur Result
Mean

Laboratory Proficiency Testing in the Australian Wine Industry – A Cooperative Model

S. M. Weeks and N. G. C. Bruer

THE AUSTRALIAN WINE RESEARCH INSTITUTE, WAITE ROAD, URRBRAE, SOUTH AUSTRALIA 5064, AUSTRALIA

1 INTRODUCTION

Proficiency testing is of little value unless it is supported by appropriate procedures for subsequent corrective action. This paper discusses the functions and activities of the proficiency testing organisation of the Australian wine industry, the Interwinery Analysis Group Incorporated. The major points of focus are the techniques used for statistical evaluation of results of analyses, the procedures for corrective action, and results achieved from this action.

The Interwinery Analysis Group Incorporated was founded in the Barossa Valley of South Australia in 1983, by several winery personnel who saw the benefits of cooperation with other, similar laboratories. The Group has grown substantially since that time and now boasts over 65 members, who together account for over 95% of Australia's wine production. Every major winemaking region in Australia is represented. The Group is administered by elected members from both wineries and allied organisations, such as independent government and wine industry analysts. Nine of the members are registered with the National Association of Testing Authorities (NATA) in the field of chemical testing.

2 OBJECTIVES OF THE GROUP

The Interwinery Analysis Group Incorporated has four major objectives, all of equal importance to its continued success. These objectives are as follows:

- to establish a high level of proficiency in the field of wine analysis by regular cross analysis and evaluation of results;
- to encourage investigation and research in all aspects of analytical methods relevant to wine and wine related products;
- to secure for members the advantage of cooperation and unity of action; and
- to hold regular meetings and generally encourage the acquisition and dissemination of useful information relating to wine and wine related products.

3 THE TESTING PROGRAM

The comprehensive proficiency testing program includes sixteen of the most common wine analyses in each of the six rounds of testing, conducted annually. Two samples of similar composition are distributed for analysis in each round. The committee purchases the wines for each round to ensure that a diverse range of wines of differing style and quality is analysed over a period. For example, a dry red wine may be chosen for one round while a sweet white wine might be chosen for another. The analytical results from each participant are collated prior to each meeting and statistically analysed using the Youden graphical diagnosis[1]. The results and related issues are discussed in detail at a meeting held after every round of testing. Guest technical speakers also address the meeting on relevant topics such as new processes and research programmes.

3.1 Youden Graphical Diagnosis

This particular method of statistical diagnosis is a valuable tool, as it provides a simple, visual method for the assessment of the precision of the results obtained by the participating laboratories. An example of the results of the Group for a given analysis as described by the Youden diagnosis is shown in Figure 1. The results for sample A are plotted on the x-axis, while the results for sample B are plotted on the y-axis. Two lines representing each of the means of these results intersect at the centre of a circle; this circle represents the 95% confidence interval. It is the aim of members to achieve analytical results which lie within this 95% confidence interval. Results obtained by a laboratory which occur outside this confidence interval but still lie on or close to the 45 degree line (shown as O) usually indicate a systematic error. The laboratory may, for example, be obtaining consistently low results for a particular analysis, in comparison with the results of the Group overall. This could be due to such factors as incorrectly standardised reagents or glassware, which may give rise to results which are reproducible, but not accurate. Errors that do not lie either in the 95% confidence interval or on the 45 degree line are considered to be random errors, as there is poor correlation between the results for each sample (shown as □). The further each pair of results lies from the diagonal, the greater the error. Errors of this nature may be caused by factors such as operator error or equipment malfunction.

3.2 Results

The results of the Youden diagnosis are circulated to each member prior to the meeting, so that their agreement with other members of the Group can be determined for each analysis, by observing their position on the graph. The results for each laboratory on the Youden diagram remain confidential, indicated only by a black square. If two or more laboratories obtain the same pair of results, the number of laboratories sharing that point is indicated on the diagram.

If the Youden diagnosis shows that several laboratories have obtained results which lie on or about a single position which is outside the 95% confidence interval, appropriate corrective action will be discussed at the meeting. Such variation may be caused by the modification of existing, or use of alternate, analytical methods.

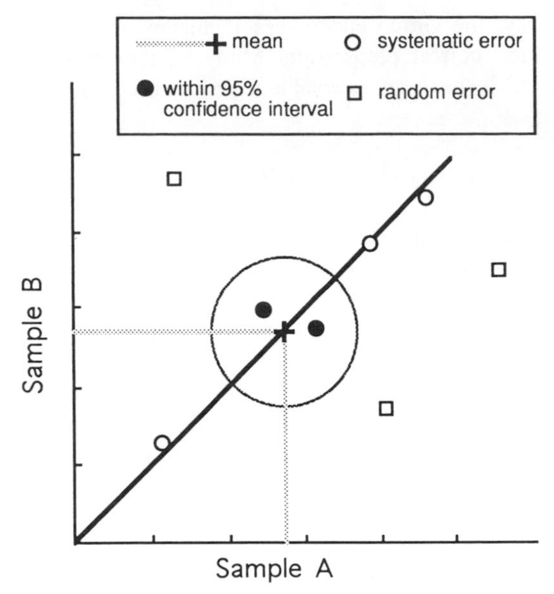

Figure 1 *The Youden Graphical Diagnosis: example of scatter plot generated from one round of analysis*

3.3 Corrective Action

Corrective action in response to poor results is the responsibility of each individual laboratory. Test samples from the previous round are available for those members with outlying analysis results, to check the effectiveness of their corrective action. If a large spread of systematic errors occurs in any one analysis, it is brought to the attention of the Group at the meeting. The discussion of results is an agenda item at each meeting, which provides an opportunity for staff of these laboratories to seek advice from other technical personnel who may have experienced similar difficulties in the past. This forum also provides opportunities for the general discussion of trends and other analytical discrepancies and/or difficulties. Possible sources of error are discussed, including methods, equipment calibration, and differences in operator technique. A plan of action is minuted and carried out over subsequent meetings until results are satisfactory.

Following group discussions, specific aspects of an analytical procedure (for example the reagent concentration or sample volume) may be agreed upon. These are distributed to Group members for comparison with current methods, and adoption where appropriate. Standardised solutions are also issued from time to time, as required. Those members who cannot attend meetings receive copies of these methods, as well as regular newsletters outlining the activities of the Group and the formal minutes of the meeting.

3.4 Limitations of the diagnosis

The Youden graphical diagnosis does have limitations. The composition of the two samples analysed must be similar for the subsequent diagnosis to be valid. This method of statistical analysis tests the precision of the results provided by the laboratories, rather than accuracy.

To assist with accuracy, the Group encourages inclusion of appropriate standards and offers methods for their correct preparation. Many of the laboratories have standard recovery tolerances, which ensure an acceptable degree of accuracy.

4 IMPROVEMENTS BROUGHT ABOUT BY THE ACTIVITIES OF THE GROUP

The Group has achieved a great deal since its inception. Several achievements, which typify the way in which the Group strives to continually improve techniques for wine analysis, are discussed below.

4.1 Determination of Volatile Acidity

Concerns were expressed at successive meetings regarding results obtained for the determination of volatile acidity. As a result, a survey of the methods was carried out, and considerable differences in procedures were identified between laboratories. Participants were provided with an aqueous acetic acid standard solution at a subsequent meeting, in order to eliminate problems with incomplete recovery and/or improperly standardised titrants. Sample volumes and titration indicators were also scrutinised, and the most appropriate were suggested for future work. The precision of the results obtained by members of the Group improved to some extent, but discrepancies amongst some laboratories persisted.

Part of the procedure involves the addition of hydrogen peroxide to prevent the interference of sulfur dioxide which may be present in the wine. Trials were carried out at The Australian Wine Research Institute, and it was discovered that the amount and/or concentration of hydrogen peroxide added by some laboratories was far in excess of that required to oxidise the typical concentration of sulfur dioxide in wine to sulfuric acid. Analysis by high performance liquid chromatography (HPLC) indicated that excess hydrogen peroxide was oxidising other wine components to volatile acids, thus giving erroneously high results.

The results of the Institute's trials were presented to the Group and compared with the results of similar trials done by other laboratories. A suitable concentration of hydrogen peroxide to be added was determined, and recommended for adoption. The precision of the results obtained by laboratories following these investigations has improved substantially. It is anticipated that statistical analysis of these data will in due course demonstrate the significance of this difference.

4.2 Alcohol Analysis

Another of the Group's major achievements is the development of an accurate technique for the calibration of alcohol hydrometers, used for the determination of alcoholic strength of wines and spirits. The analysis of alcoholic strength has always been a contentious issue in the Australian wine industry, due to the stringent requirements of export documentation, customs and excise. The several methods used (distillation/pycnometry, distillation/hydrometry, ebulliometry, HPLC, gas chromatography and distillation/dichromate oxidation) have given different results, often varying by as much as 2% v/v. Specific aspects of each of the methods used were discussed with limited improvement in the subsequent results.

The Group purchased an aqueous ethanol solution of certified alcoholic strength from a NATA registered laboratory, in order to evaluate reference hydrometers. Members brought their hydrometers to a general meeting for evaluation in this solution. The temperature of the standard was maintained at 20°C to eliminate the need to apply temperature correction factors. It was revealed that seventeen of the nineteen hydrometers checked gave results which were 0.3–0.8% v/v below the strength of the certified solution. In response, the Group asked that The Australian Wine Research Institute collaborate with the organisation that had calibrated the hydrometers, to identify the cause or causes of the apparent discrepancies. Despite discussion and examination of the procedures, none were identified. With the support of the Group, The Institute developed a method for the calibration of alcohol hydrometers based on the European Union reference method of pycnometry. This method was subsequently accredited by NATA, in order to satisfy the Australian Customs Service and other regulatory bodies. This project was completed in June 1994.

Following the use of these hydrometers by the industry, the range of alcoholic strength results reported by Group members has decreased by approximately 0.5% v/v. When one considers that the labelling specification for wine entering the European Union is ± 0.5% v/v, it is easy to recognise the significance of this achievement. The range is expected to decrease even further as use of the hydrometers calibrated by The Institute becomes more widespread.

4.3 Other achievements

The wine industry has seen many other achievements brought about by the activities of the Interwinery Analysis Group Incorporated. Among these are:

- the standardisation of methods for microbiological analysis of wines and related products, an ongoing project undertaken by a subcommittee of the Group;
- cooperation with NATA and liaison with regulatory authorities to ensure the integrity of export documentation; and
- detection of a problem with the quality of widely used disposable cuvettes, which was evaluated by members of the Group, before notification of the industry at large.

5 CONCLUSION

In summary, efficient proficiency testing requires:

- a clear procedure for the determination of non conformance;
- corrective action procedures;
- opportunities to repeat analysis using the same samples; and
- opportunities to discuss non conformances with others in an sympathetic, non threatening environment.

The Australian wine industry has demonstrated its commitment to quality analytical results by bringing together qualified and experienced technical staff to pursue a common goal, despite the competitive nature that might be expected in an industry of this type. The quality of this analysis has contributed to the high standard of Australian wine in both domestic and international markets and we are very proud to be a part of it.

Acknowledgment

Many thanks to Dr Bob Dambergs of BRL Hardy for his help and support during the writing of this paper.

Reference

1.W. J. Youden, ‘Precision Measurement & Calibration, N.B.S. papers on statistical concepts and procedures’, U.S. Dept of Commerce, Washington D.C., 1969, p133-137.

Laboratory Certification to ISO 9000: 2.5 Years Experience

P. H. Wright

GAS GRID QUALITY ASSURANCE BRANCH, ENGINEERING SERVICES DEPARTMENT, GAS & FUEL, 1136 NEPEAN HIGHWAY, HIGHETT, VICTORIA 3190, AUSTRALIA

1 INTRODUCTION

ISO 9000 quality management systems provide the foundation around which TQM improvements can be made and measured.

The Gas & Fuel Scientific and Engineering Services Laboratory complex gained certification to ISO 9001 in March 1993 (QAS certification QEC 1454). The two and a half years experience gained since the certification has lead to improvements in the management of the various sections of the laboratory.

1.1 The Gas & Fuel Laboratories

The Gas & Fuel, which supplies gas throughout the state of Victoria has a laboratory complex in the Melbourne suburb of Highett. The laboratory consists of a number of branches; Gas Quality & Environment, Appliance and Gas Utilisation branches under Scientific Services Department and Gas Grid Quality Assurance branch under Engineering Services Department. The diversity of functions carried out by the different branches and the different reporting structures required a flexible quality management system.

1.2 The ISO 9001 Accreditation

The current accreditation covers chemical testing and evaluation, analysis of fuel and industrial gases, calibration and measurement of gas flow, pressure and density, gas appliance research, development, testing and evaluation, energy management modelling, design and consulting, materials testing and evaluation, various NATA accredited testing services in the fields of Heat and Temperature, Chemical and Mechanical Testing and many related services.

2. MANAGEMENT UNDER ISO 9001

An important point to be made is that the ISO 9001 quality management system is not a separate entity or an additional component of management. The ISO 9001 certified system is *the* management control system. It is also important to note that although

ISO 9001 mandates *what* must be controlled, management has full discretion as to how to achieve that control.

2.1 The Early Days

The task involved in gaining a quality management system certification depends on how controlled and documented the existing management system is. In the Gas & Fuel laboratories although many components of the management system were in place they were not integrated and were of varying levels of completeness and control.

Implementation of ISO 9001 initially resulted in unrealistically high requirements for documents and document control. This arose because there was a natural desire to attempt to put into place an "ideal" system. This unfortunately created a "paper dragon" with too many forms and too much detail in procedures, the emphasis had been too much on the *system* not on *management* of the output.

A wonderfully documented quality management system with lots of forms requiring lots of signatures may fulfil all of the requirements of ISO 9001 but will not necessarily help the management of the organisation if it is so cumbersome as to be unworkable.

2.2 Experience Gained

One of the important lessons learned was that a quality management system put into place should never be regarded as fixed for all time. Circumstances, business goals and management structures all change and the management system must not only be able to adapt but should actually assist in responding to external changes.

It was possible to simplify and improve the system as time went on. Part of the streamlining of the system came about from the realisation that the quality system should not be an additional management task but should be a tool to be used. Our special ISO 9001 "quality management" meetings became the normal management meetings.

It also became very clear that the system had to be flexible, to allow for constant change. Cross referencing between various quality manuals, procedures and work instructions was reduced to a bare minimum so that the removal or significant change of a document did not require a flow on of changes in related documents. Individual procedures and work instructions were considerably reduced in size to only include what was necessary to reduce the need for document changes for inconsequential details.

As part of allowing for constant change, work instructions were kept very task specific. It was found to be preferable to have five specific work instructions covering five related tasks than to try to make one all-encompassing document or manual. Smaller documents were easier to read and use. If changes were necessary, the task of preparing new small documents is not seen as so daunting as to upgrade a thick manual.

3 INTERRELATIONSHIP OF ISO 9001 AND TQM

ISO 9001 is essentially a model for business functions management in that it lays down those essential functions that must be controlled to run a business. It should thus be regarded as the minimum required for good management. The standard does not address issues such as human relations, staff involvement, process improvement, marketing or many other issues which are essential elements of quality management. What ISO 9001 does provide however, is a system where vagueness is not tolerated and functions and processes are defined. The very act of defining processes provides the starting point for TQM type improvements.

ISO 9001 can easily be implemented in an organisation without embracing TQM. Similarly TQM could be implemented without a certified ISO 9000 type quality management system but the full benefits may be more difficult to realise. An ISO 9000 system provides a way of implementing and maintaining improvements developed through TQM.

The Gas & Fuel policy is a commitment to the Australian Quality Award's categories and guidelines to focus its improvements and has adopted the ISO 9000 series of standards as the foundation for quality improvement.

4 COSTS OF MAINTAINING ISO 9001 SYSTEM

The CSIRO has estimated[1] that the costs of TQM for an organisation to be approximately 2.5% of turnover. In a laboratory environment turnover translates loosely into running costs.

The costs of maintaining the ISO 9001 certification within the Scientific Services laboratory are 0.5% of running costs, essentially the cost of about one third of one staff member spent as a document controller/audit co-ordinator.

The preparation of new documents, updating existing documents, and carrying out actions required by the ISO 9001 standard are not classed as a cost of maintaining the ISO 9001 certification. This statement may seem strange but it was one of the realisations made in operating a certified quality management system that the above tasks were what should be done as part of normal operations in a well managed organisation. The documentation tasks are not an additional impost and should not be separated from normal operations.

In the same manner the costs of implementing an ISO 9001 quality management system should not be seen as a "cost of certification" but as a cost of "getting the business under control". If a business is well managed (good documentation is seen as part of that) then gaining certification is a relatively simple task. If it takes a great deal of effort to put a quality management system in place then it must be appreciated that a significant portion of this cost is most likely a catch up from the "years of neglect".

5 BENEFITS OF ISO 9001 SYSTEM

An ISO 9001 quality management system imposes a discipline on all levels of management and staff. In many ways it ensures that many of those "important but not urgent" tasks get done. Often tasks which fall into that category are related to strategic management, staff training, records, and documenting work methods, and these are the very issues which are part of good management.

Knowing that the management system is going to be audited even if only by your own staff as part of internal audits does mean that documents are kept up to date and that things which should be done, actually are done.

Issues as to who is responsible for what are defined, methods are documented, appropriate records are kept and, in other words, good management practices are promoted.

The most tangible benefit that has been identified that has had the greatest overall effect has been the improvement in customer focus. As part of a laboratory function of a significant gas utility the only customer was previously seen as "the utility", focussing on quality management of the laboratory has highlighted the many diverse "customers" within the utility, each with their own individual requirements and expectations.

Quality management is very much about satisfying customer needs. ISO 9001 provides the basis for analysing ones business, provides the hard data necessary for improving ones business and provides the vehicle for enshrining improvements into the day to day operations of a business all of which feed the TQM improvement cycle.

ISO 9001 certification has a strong effect on staff involvement in process improvement. A documented management system is a visible indication of the current authorities, responsibilities and processes and so process improvements can be openly compared with the existing systems. Process improvements, once made, are clearly visible to all, so staff can see concrete evidence of their efforts and more importantly can see that they *can* make a difference.

6 CONCLUSION

ISO 9001 certification provides the "hard facts" core of quality management which in turn provides the base upon which to build changes in corporate culture and system improvements which are part of the essential elements of TQM. Without some recognised, documented starting point, improvement is difficult to initiate and still more difficult to implement.

The most important lesson learned over the last two and a half years may appear obvious but cannot be stressed too strongly namely - keep it simple!

7 ACKNOWLEDGMENTS

The author wishes to thank the Gas & Fuel for permission to publish this paper.

The opinions expressed in the paper are those of the author and do not necessarily represent those of the Gas & Fuel.

References

1. CSIRO, *CSIRO Manufacturing Month,* March 1992.

Improving Analytical Quality in Clinical Laboratories Using Low-cost Control Procedures

Adam Uldall,[1] Maritta Siloaho,[2] and Eino Puhakainen[2]

[1] DEPARTMENT OF CLINICAL CHEMISTRY, HERLEV UNIVERSITY HOSPITAL, DK 2730 HERLEV, DENMARK

[2] DEPARTMENT OF CLINICAL CHEMISTRY, KUOPIO UNIVERSITY HOSPITAL, FIN 70211 KUOPIO, FINLAND

1 PREFACE

1.1 Improvements of analytical quality in the perspective of total quality assurance in clinical laboratories in geographic areas with limited resources

Supporting quality assurance (QA) in clinical laboratories may involve organizational elements, educational activities, and external quality assurance programmes. Such activities are clearly intended to be long run projects, typically needing more than five years to become completed - if that will ever happen. Ideally, means should be made available in each laboratory to build up proper quality systems, reliable routine measurement procedures as well as proper internal quality control procedures. Obviously that would in many areas make high demands on the resources. However, steps in the proper direction could be achieved through educational activity. Priorities should be given to each element of a QA programme so that all steps are done in a proper sequence, sufficiently slow for the work to be practicable, and so that success is achieved at each step.

Clearly, the QA-activities should be driven by colleagues in the involved areas and the most important contribution from abroad would be to establish courses or just open existing QA courses in their own countries for local key persons. Lectures given on national or regional level in the relevant areas may also add to the improvements.

At the organizational level in the relevant geographic area both supranational, national, and regional scientific societies for laboratory medicine should form QA-committees. These committees should establish working groups on various aspects of quality assurance e.g. education in QA; proper nomenclature; guidelines on proper test selection; proper specimen collection, transport and pretreatment; selection of proper measurement procedures; internal quality assurance (including certain aspects of GLP), internal quality control, external quality assessment schemes; reference ranges; proper reports to the clinicians; utilization of EDP etc. Section 6 of this paper is a basic example of how to run an analytical QA programme step by step.

This paper provides some general guidelines; they could be considered as goals in many situations; furthermore tools for improvement of the analytical quality may be extracted. The instructions, practically derived from this paper, should be written for the laboratories in their own language.

Obviously, each society cannot cover all aspects at the same time, therefore close cooperation between various societies near by may help.

1.2 The FESCC QA guideline project

Forum for European Societies of Clinical Chemistry (FESCC, "IFCC in Europe") has initiated and endorsed the below indicated QA guideline project concentrating on pure analytical problems. It is proposed to issue guidelines on achievable implementation of internal quality control (IQC) and external quality assessment schemes (EQA schemes) under the current practical conditions in certain parts of Europe. To edit such guidelines, formation of an FESCC working group (FESCC WG) is proposed. This working group shall extract relevant matters from already existing documents. From this the group will compile an FESCC guideline on analytical QA in clinical chemistry laboratories in relevant parts of the formerly Soviet dominated Central Europe, and when possible, the corresponding parts of Eastern Europe. One source for this work is the IFCC-draft paper on EQAS (1). Another source is the internal quality assurance paper presented below; an attempt for a more practically oriented instruction is presented in a brief form in section 7 of this paper. Plenty of further documents should be considered, e.g. ISO guides (2), ECCLS-IFCC guidelines for decentralized testing (3), IFCC guidelines in quality control, guides from WHO-EMRO (4), WHO-SEARO, and WHO-PAHO, and commercially available instructions. But most important of all will probably be to identify and possibly translate useful papers already in use in some of the relevant areas.

The project will be established in collaboration with IFCC and other relevant organizations. The FESCC WG involves in principle colleagues from all relevant countries. However, for practical reasons the FESCC WG should only include a small number of individuals from some of the countries where quality problems are realized but far from being solved yet.

2 INTRODUCTION

The objective of this document is to provide suggestions for establishment of a low-cost improvement programme for analytical quality in clinical laboratories located in areas with only modest resources available. It is designed to assist clinical chemists in the practice of analytical quality assurance including internal quality control. The basic concept of the present approach is firstly to encourage establishment of what might be called proper laboratory practices (6) and secondly to support that by external quality assessment of the internal control results. Low-cost internal control procedures using patient-samples and patients' results are used on a day-to-day basis as far as possible to monitor the analytical performance. Many of the internal control procedures may be facilitated by some kind of laboratory EDP-system to collect the patient data and perform the calculations. In addition, conventional EQA (proficiency testing) is necessary (1).

3 DESIGN OF A PROGRAMME FOR INTERNAL ASSURANCE OF ANALYTICAL QUALITY BASED ON PROPER LABORATORY PRACTICES

3.1 Basic requirements of proper laboratory practice to be implemented in the laboratory

The below topics for good laboratory practices are selected from a larger document, dealing with quality systems in clinical laboratory (5) and Scandinavian recommendations on quality assurance (6). These documents should be consulted for more details. The present use of the concept "proper laboratory practices" should merely be understood just as the words say - it does not refer to specific requirements, in contrast to "Good Laboratory Practice" in the OECD sense.

The main elements of proper laboratory practice are:
(a) ensure necessary facilities, well educated staff and quality system; (b) establish quality specifications for the laboratory service; select analytical methods which can fulfil the specifications; do proper documentation of the selected methods and instrumentation including criteria for the choice, evaluation report, descriptions (measurement procedures), oral instruction etc. Keep a logbook of major instruments, do preventative maintenance, do control check of instruments etc.; (c) ensure safe instruments, safe procedures for the use of reagents, utensils and instruments; (d) ensure proper sample collection and handling; (e) establish sufficient quality control procedures - and provide guidelines for trouble shooting; (f) participate in EQA schemes; (g) monitor the validity of the currently used reference intervals; (h) ensure proper communication with the clinicians.

Some basic elements to start with for the proper performance of a laboratory may include:
- organized technical service for the laboratory instruments;
- facilities to produce and store purified water;
- safety instructions for the laboratory including instructions for handling and disposal of infected materials, dangerous chemicals, and waste.

Below are presented some basic good practices and those quality control matters which may not require too much acquisition.

3.1.1 The laboratory should select analytical methods which are capable of fulfilling the level of needed analytical quality as far as possible. Obviously the selected measurement procedure should be as stable as possible, so that within-run drift is minimized and variation between the analytical runs is small; achieving calibration stability over long periods of time is a further essential advantage. The random analytical variation (s_{random}) of the final result should be explainable by the laboratory as to the accumulation of the variation (s_n) of each step according to the formula: $s_{random} = \sqrt{s_1^2 + s^2 + ... s_n^2}$.

3.1.2 The laboratory should select, use, and maintain analytical systems or instruments, reagents and utensils in accordance with the description provided by the manufacturer or local professional authoritative bodies, or well documented published method. Log book for instruments showing the maintenance performed, malfunctioning, repair, and control should be kept.

3.1.3 The measurement procedure should be calibrated with utmost care involving multiple determinations, ideally by using materials which are traceable to primary standards. If, e.g., the analytical variation is 1 and the calibration is done once only, the total variation of a single results is $\sqrt{1^2 + 1^2} = 1.4$. If calibration measurements are done four times, the total variation of a single result is reduced to $\sqrt{1^2 + (1/4)^2} = 1.1$. For simplicity, this example is selected to illustrate the situation when calibration at only one concentration is needed (besides the zero concentration where the uncertainty is ignored): for measurement procedures where a calibration function needs to be established experimentally more calibration "points" are needed.

3.1.4 The laboratory should be able to present documents showing the background for the selection of the method and its local approval.

3.1.5 The laboratory should provide and continuously update detailed descriptions of how to do the measurement: the measurement procedure. An illustrative short version suitable at bench level may be useful as well cf. (5). All instructions and changes should be dated and carry the signature(s) of the responsible person(s).

3.1.6 The laboratory should instruct orally the individual technician about how to perform the particular measurement. The new procedure should be carried out several times by the technician before doing it as a routine; and after appropriate exercise, comparison of obtained data with findings on the same patients' - and control samples by that of more experienced technicians should be done. When the laboratory management has approved the comparability of the results this should be recorded and stored in the laboratory file. A given measurement procedure may only be carried out by technicians who locally have passed that approval.

3.1.7 Dated original measurement results on patients' and control samples together with their identification should be signed by the analyst and should be stored for an appropriate time.

3.1.8 Before release of results certain control procedures issued by the laboratory management have to be passed. They may comprise reliability judgments (3.2), use of analytical control rules (4.1, 4.2) check of unusual dilution of the sample, independ recalculation of results, and check of correct transfer to the report. The analytical control rules depend on arrangements concerning the internal quality control and may include:
- pools;
- patient samples;
- commercial controls;
- control charts;
- statistical procedures requiring PC or/and laboratory EDP.

3.1.9 The laboratory management should appoint individuals who are responsible for the daily quality control of each measurement procedure and those who are responsible for long term quality control, so that all data are evaluated regularly - possible monthly.

3.1.10 The laboratory should describe in details how patients' specimens should be taken, stabilized, transported, pretreated and stored. Permissible storage temperature and acceptable storage time of samples should be based on knowledge of the stability of the material. Strong attention should be paid to the patient-identification and labelling of the primary specimen as well as further laboratory samples obtained from the specimens.

3.1.11 Regular inspections initiated by the laboratory management should verify the appropriateness of all procedures. Record keeping of such inspection as well as possible corrective actions should be done.

3.1.12 A representative from the laboratory - often the director - should on a regular basis meet the doctors or groups of doctors and nurses of each ward, served by the laboratory in order to identify possible deficiencies and to find out how the analytical service could be improved. Record keeping of such meetings and of possible corrective actions done is essential.

3.1.13 Regular revision of the reference ranges provided by the laboratory; that may be established in cooperation with other laboratories in the region.

3.2 Judgement based on distribution of patients' results

3.2.1 The distribution of patients' results depends on the clinical situation, blood collection procedures, preparation, transport and storage of sample as well as the analytical quality. Consequently no firm analytical decisions can be taken solely on the distribution of patients results. However, important information may be observed because most requested quantities notoriously are "normal" and because many hospitals and clinics have a fairly stable distribution of different clinical cases. Therefore record on a control chart weekly or monthly of the median patient value and the fractions of patients' results which are above upper reference limit and below the lower reference limit is an important aid. The median value should be located close to the central part of the reference interval, while the fractions of results outside the reference limits usually is between 5-15%, but depend on the clinical situation. Larger changes in the median value or the fractiles than caused by statistical phenomena could be caused by preanalytical conditions e.g. contaminated blood sample tubes. Having regular recording of these data an experienced colleague may judge roughly about the appropriate reliability of results of each run of analysis with only 10-20 patients' specimens. However, such approvals are only provisional because control sera are indispensable for a scientifically correct approval. A laboratory may for economical reasons not do control sample measurement in each run but observe e.g. the location of the patient median value. In case of a changed median value, the laboratory should then, in order to exclude analytical errors, introduce control measurements using control sera with known value.

3.2.2 Often, the laboratory knows earlier findings on the same patient, in which case comparisons may support approval of the actual run if several sets of such results are available; however, often also deviating results are found because of the therapy. Therefore close contact with clinicians is needed when utilizing this type of information. Furthermore, such comparisons based on few data sets may incidentally provide either false alarm or false approval.

4 INTERNAL QUALITY CONTROL BASED ON CONTROL MEASUREMENT AND CONTROL RULES

4.1 Strict rules based on measurement on patients' samples

4.1.1 Often, measurements on patients' samples are done twice in the same run. This is suitable to monitor the part of the random variation caused by the within-run variation. A within-run standard deviation for these duplicates can be calculated using the formula $s = \sqrt{\Sigma(d)^2/N}$, where d is the difference between the two readings, and N is the number of duplicates. This standard deviation may be compared with a similar standard deviation obtained when the method was run under optimal conditions. However, such standard deviations are only valid when calculated from specimens within a narrow concentration range, because the analytical standard deviation usually increases with concentration. (A useful way to overcome this problem is to establish an "imprecision profile" over the whole measuring range taken from each result of the double determinations performed on the patients' samples when the method was well functioning; this profile is then compared to similar data obtained in the actual analytical run for approval). It should, furthermore, be noted that the way the duplicates are performed influences the s, e.g. whether each single determination of the duplicates are measured immediately after each other, or different samples are measured in between. Therefore, the s should be calculated from duplicates which are run in the same way as used in routine determinations.

4.1.2 A control rule for approval of the within-run variation of a run could be that if the single determinations of a duplicate deviates more than e.g. $2s$ the result is rejected* and if more duplicates exceed that limit the whole run is rejected; obviously the acceptable number of sampling with deviating results depends on the length of the analytical run. Remedial actions see 4.3.

4.1.3 The same principles as presented in 4.1.1-4.1.2 can be used to monitor the between-run variation when determining more patients' retained samples ("repeatends") from a previous run in the next run. The between-runs s of duplicates could be determined accordingly. Stability of the sample material between the runs is an essential prerequisite. When storing human sera in stoppered glass-tubes at +4 °C in darkness for 2-3 days this applies to most components in the basic clinical chemistry menu (for glucose and total CO_2 special precautions are needed); however, it cannot be excluded that some analytes may show apparent instability. The recommended control rule (7) when using retained patients' samples is: runs where two of n samples show a deviation of $\geq 2\,s$ at the second determination compared to the first determination are rejected. The probability of false rejection and the size of the systematic error to be detected (with a probability of 0.50; P_{ed} = 0.50) are shown in Table 1.

Table 1. *The suitability of control based on retained patients' samples*

Number of retained samples n	Probatility of false rejection P_{fr}	Systematic error, detectable with the probability P_{ed} = 0.50
2	0.00	3.2 s
3	0.02	2.4 s
4	0.04	2.1 s
5	0.06	1.8 s

*Unfortunately no generally accepted power functions of such rules exist yet; however, for the rules in 4.1.3 documentation exist when used in hematology.

4.1.4 The approaches in 4.1.2-4.1.3 are good supplements to the judgments on distribution of patients' results when control sera cannot be measured in each run. However, the control of systematic error is weak and the power of error detection is not too good; never the less, these approaches are indispensable when no control sera are available for all runs. It should be remembered that control with known control materials should be performed at least now and then when apparantly needed.

4.2 Strict rules based on measurement on control sera

4.2.1 Storage of the lyophilized control serum should take place in a refrigerator. For longterm storage even -20 °C is advantageous. See also 7 Annex B.

4.2.2 A good approach will be if at least two determinations of the control material are done in each run and at least five consecutive runs evaluated according to the rules described in 4.2.3 (if the patients' samples are reported as mean values of duplicates, each of the two control measurements should also be the mean of duplicates). This will provide a high rate of properly validated runs at the time of delivery of results.

4.2.3 The following set of control rules could be selected, e.g.:
The number of previous runs covering a total of 10 determinations of the control serum should be looked at. An alert is present if:
- one out of the last control results is outside the ± 3*s* limits;
- two consecutive control results are outside the same ± 2*s* limit;
- the range between two consecutive control determinations exceeds 4*s;*
- four of the previous control results exceed the same 1*s* limit;
- all last 10 control results are found at one side of the target value.

The first three rules should be used as criteria of rejection of the run. The two last rules indicate needs for maintenance.

4.3 Trouble shooting

4.3.1 A guide of trouble shooting should be issued. The type of error indicates frequently a selection of possible remedial action. The type of errors observed in internal quality control are usually classified as <u>random</u> or <u>systematic</u>.

4.3.2 <u>Random errors</u> manifest themselves by variations in the result of repeated analyses of the same sample. These variations can be caused by more or less independent factors such as variations in apparatus, temperature, electric current, weighing, pipetting etc. Random errors are unavoidable; they will always be present to some extend and will influence results. Often random errors are expressed as standard deviations. In general, even though random errors cannot be eliminated, they can often be reduced e.g. by using more stable technique.

4.3.3 <u>Systematic errors</u> can be due to the analytical method, technical performance, the reagents, the measuring equipment, the calibration and the actual measurement procedure. It is characteristic of a systematic error that it can always be traced back to a definite cause which may be isolated and eliminated, e.g. erroneous calibration, wrong reagent, wrong pipet. Some errors provide a constant deviation. Others provide a deviation which is dependen on the concentration e.g.

a certain pipet with a wrong adjustment. Basicly, one should always correct for a systematic error when it is known either by removal of the cause or by applying a mathematical correction; however mathematical correction is not much used because it requires that the error is stable over time and that one is convinced that the error is real; finally it often requires the use of highly elaborated reference methods.

4.3.4 Signals from the control system indicate the type of errors. Having found e.g. a deviating median of patients results, or having found all 10 consecutive control results higher or lower than the target value, this indicates a systematic error. Then one should look at the calibration including the shape of the calibration curve, the adequate composition of reagents, the wavelength etc. Having too large discrepancies between repeated measurements e.g. the repeatends or one control observation outside 3*s* limits indicate a random error. In this case one should look e.g for unstable reading, unstable pipetting, poor mixing etc.

4.4 Trueness control

Quantitative expressions of trueness may be presented as an inverse term, "bias" or in a more colloquial style "systematic error". Basicly the systematic error should be negligible, however, in practice some systematic error is unavoidable. To balance the types of errors one may accept as a rule of thumb an ignorable systematic error of less than one half to one CV of the attainable between run variation, but of course that depends on the clinical use of the results.

The best way to control trueness is to analyse native samples that were previously analysed in a reference laboratory. The prerequisit is that the laboratories are located rather near each other and have proper sample transport facilities.

The use of a certified reference material is a useful tool to establish the trueness of the measurement procedure when a reference method is not available. A reliable control material used on a daily base and by other laboratories in the area may substitute for the expensive certified reference material. Mean and standard deviation are calculated from the results of the control material when e.g. 6-20 analytical runs with control results are done; such calculations should be carried out on a regular basis. The findings are compared to the target value relevant to the locally used measurement procedure established in other laboratories or earlier in own laboratory. For further information on the trueness control consult clause 5.

5 MODEST-COST EXTERNAL TRUENESS ASSESSMENT USING COORDINATED PROVISION AND USE OF INTERNAL QUALITY CONTROL SERUM COMBINED WITH INTERLABORATORY DATA COMPARISON

5.1 Operation of the control

On a regional basis a group of laboratories should procure or reserve for later stepwise procurement a certain batch of control serum for internal quality control purpose e.g. for two years of use. Alternatively, local production of the material could be performed (1). The control data collected in each laboratory over a specified period of time serve as a relative trueness control when results from all laboratories are compared - or a more absolute trueness control when the comparison is done to a reliable value. Several regions have had benefit of this approach (e.g. 8, 9, 10). The specifications of the serum can be found in (1) - the serum may be of human or animal origin. The serum is used within each laboratory as described in 4.2-4.3. Monthly or quarter annually, means and standard deviations and other essential information is reported to a coordinating laboratory which produces written surveys. Usually the results are separated into methods groups. It would be useful if results obtained on patients' samples in the same period of time be compared e.g. so that patients' medians are shown together with the mean values on the control serum. Agreement between deviating control data and deviating patients' medians indicates a need for remedial actions in the analytical field; disagreement between these parameters may indicate the need for further studies, e.g. more patients' data, preanalytical factors or another type of control serum. Also the locally used reference intervals could be included in such studies.

5.2 Assigned values to the control serum for trueness control

Reference measurement procedure approach would be most ideal, but seldom available; however, reliable routine measurement procedures calibrated with certified reference sera are more economically achievable. Another approach is to use "qualified" consensus value.

If only few laboratories participate into the trueness control, the serum may need to have assigned values from an external quality assessment scheme - possibly done elsewhere. When many laboratories participate in the trueness control they may provide the consensus value themselves. When the control serum has been circulated in an external quality control scheme including some laboratories with higher metrological experience, a judgement of a proper target value may be given. The external quality control scheme should include separation of results into groups of instruments and principles of measurements in order to provide a background for this judgement. Careful evaluation of the distribution of results in the external quality control scheme, combined with the methodological considerations, is important.

5.3 Suitability of the control serum for various measurement procedures.

The matrix of the control serum and the type of components added to the control serum influence the results of different measurement procedures. For several of the basic components in a clinical chemistry laboratory this has not been a major problem; however for e.g. total protein and for albumin the bovine serum may show measurement procedure dependent values (11). It is essential when using such an "arteficial" control serum to consider possible measurement procedure dependent influences on the expected values. Indications of such problems are most conveniently gained through the external quality assessment scheme where the findings are recorded according to type/or brand of instrument and principle of measurement. Having indications of such a phenomenon, confirmative comparisons are needed. Such studies could be carried out in one laboratory where the measurement procedure with the deviating findings is compared to another procedure with proper results on the control serum. The study should establish, whether the discrepancy reflect only the control serum and not the patient samples.

6 SUGGESTED PLAN FOR IMPROVEMENT OF ANALYTICAL QUALITY IN CLINICAL LABORATORIES IN A CERTAIN REGION

The below sequence of steps is an example which might be a proper choice:
- organizing a groups of clinical chemists interested in quality improvements;
- planning education, writing or translating instructions for the laboratories in their own language;
- establishment of reference laboratories with good quality reference instruments, e.g. spectrophotometers and cell counters, and set up of reliable calibration procedures;
- proper facilities for purification of water, preparation of reagents, and sample storage (refrigerator, deepfreezer) in the laboratories should be ensured;
- guidelines on how to select proper methods under the particular situation should be given; maybe recommendations on selection of certain standard method could be provided;
- the robustness of the instruments should be considered before procurement. This is important in laboratories where the supply of electric current, fluctuations in voltage, room temperature, dust or mechanical vibration may cause problems.
- technical service-procedures for the instruments, and setting up maintenance procedures are needed;
- writing working instructions for collection of samples and their preparation, for the analytical procedures etc., are essential.

7 CALCULATION OF TARGET VALUES AND ACCEPTABLE LIMITS OF CONTROLS. FORMATION OF QUALITY CONTROL CHART. An example of instructions in a brief form.

7.1 Control materials

Some general properties required of materials used as controls should be established:
- stability;
- homogeneity;
- availability in suitable aliquots;
- the same matrix ideally as the patient's sample;
- concentrations of analytes should be of suitable levels. Normal and abnormal levels are used depending on the level that is critical to the medical interpretation.

Other considerations to be taken into account:

Commercial controls:
- human: often identical matrix to the samples - but not for all individual components;
- bovine origin: cheaper, safer (no HIV or Hepatitis risk)
- assayed: target means and acceptable limits determined and stated by the manufacturer for each lot of control and often for separate analysers;
- unassayed: only the approximate level of the analytes given, cheaper than assayed material;
- when freeze-dried (lyophilised) material is used, the activity of certain enzymes, particularly alkaline phosphatase will change with time after reconstitution.

Self made pools:

- sera (no lipemic or hemolytic samples) are collected in plastic containers in the freezer at -20 °C., thawed, filtered and HIV and hepatitis tests performed before aliquoting into suitable test tubs. (NB, number of patients' samples in each pool for testing of HIV and Hepatitis, may be up to 10 samples per pool);
- most analytes are stable for several months at -20°C;
- cheap;
- possibility to control less frequently measured analytes, for which there is no commercial control available on a suitable concentration level - and price.

7.2 Reconstitution

An example of instruction for reconstitution could be:

- ensure that the lyophilised material (the "cake") is at the bottom of the vial;
- remove the metal seal;
- release the vacuum of the vial (if relevant) by slowly removing the stopper so that the lyophilised material is not blown out when the vial is opened;
- add 10.00 ml (or 5.00 ml or 1,00 ml as appropriate) distilled water at room temperature with an accurately calibrated pipette (check that nomajor water drops are found after emptying the pipet). Alternatively, a positive displacement pipet may be used;
- cap the vial and keep in the dark (bilirubin) at room temperature for 15 min.;
- mix carefully by inverting the vial 20 times, avoiding foam formation;
- ensure visually that all the lyophilised material is dissolved;
- store in a refrigerator (in darkness).

7.3 Target values and acceptable limits

Analytical measurements:

- the control material analysed at least 10 times but preferably 15-20 times using the method to be controlled;
- this is done over a period of time, that is sufficient to represent normal variation components of the method, when the method is working properly under routine conditions;
- the time should be at least some days.

Calculation of the mean standard deviation:

- mean, $\bar{x} = (\Sigma x_i)/n$
- standard deviation: $s = \sqrt{\Sigma(x_i-\bar{x})^2/n-1}$

  where n = number of measurements (values), x_i = measured values and $\bar{x}$ = mean of the values.

Elimination of the outliers:

- eliminate values exceeding the mean by more than $3s$;
- calculate the mean and s again;

The target mean and acceptable limits:
- the target mean is the mean calculated after elimination of the outliers;
- a way to express the acceptable limits (control limits) is to give them in multiples of s, e.g. mean ± $2s$ limits (warning) or mean ± $3s$ limits (rejectional).

7.4 The control chart
- use a sheet of graph paper (or computer) to construct the control chart ;
- the concentration (the observed value) on the y-axis;
- the date and/or the number of run on the x-axis;
- usually 1 month on one sheet of paper;
- draw lines at the upper and lower acceptable limits (per cent or $3s$ limits);
- plot the observed values on the paper;
- if the value is inside the limits, the patient results of the run can be reported;
- if not, don't report the patient results. Find out the cause for the deviating control value and, if needed, repeat the whole run. Check the control again.

ACKNOWLEDGEMENTS

The authors are indepted to chief chemist Martin Kjærulf Nielsen, Næstved, Denmark, for critical revision of the text. Secretary Marianne Jensen is thanked for excellent technical assistance in preparation of the manuscript.

References

1. P. Hill, A. Uldall, P. Wilding. Fundamentals in external quality assessment. IFCC, EMD, CAQ, Herlev. Draft 94-12-28; 1-49.
2. ISO REMCO N271, November 1994, Geneva, 23 pp.
3. R. Dybkaer, D. V. Martin, R. M. Rowan. *Scand. J. Clin. Lab. Invest,* 1992, **52**, suppl. 200: 1-116.
4. M. M. El Nageh, C. C. Heuck, W. Appell, J. Vandepitte, Engbaek, W. N. Gibbs. WHO EMRO, Alexandria, 1992: 1-208.
5. R. Dybkaer, R. Jordal, P. J. Jørgensen et al. *Scand. J. Clin. Lab. Invest.* 1993, **53**, suppl. 212: 60-84.
6. Nordic Committee on Quality Control (NKK). *Scand. J. Clin. Lab. Invest.* 1990, **50**, 225-7.
7. J. O. Westgard, G. S. Cembrowsky. *Eur. J. Haem.* 1990, **45** suppl. 53. 14-18.
8. J. G. Batsakis, N. S. Lawson, R. K. Gilberg. *Am. J. Pathol.* 1979, **72**, suppl. 257-9.
9. M. Blom, A. Brock, F. Christensen et al. *Scand. J. Clin. Lab. Invest.* 1984, **44**, suppl. 172, 175-8.
10. P. Grönroos, U. Hohenthal, E. Karjalainen et al. *Scand. J. Clin. Lab. Invest.* 1994, **44**, suppl. 171, 179-86.
11. A. Uldall. *Scand. J. Clin. Lab. Invest.* 1987, **47**, suppl. 187: 1-92 (see also *Ann Ist Super. Sanitá*, 1991, **27**, 411-18).

ISO-9000: Working toward Global Quality

David L. Berner

AOCS, 1608 BROADMOOR DRIVE, CHAMPAIGN, ILLINOIS 61826-3489, USA

Evidence of commitment to quality will not only be a condition of doing business, but a requirement for survival in the growing international market place.

One of the ways of demonstrating quality products and services is by registering corporate/manufacturing facilities for participation in the International Organization for Standardization (ISO) 9000 program. Over a half-million facilities are appropriate for registration through ISO in the quality assurance program generally referred to as the "ISO 9000 Series", or "ISO 9000". So far, approximately several thousand United States (US) companies are registered, and more are participating in the ISO 9000 certification program as they discover its international importance. One has to travel only about 10 miles east from Champaign-Urbana, Illinois into a vast Mid-west farming area to find a seed company displaying a large sign: "An ISO 9002 Company".

The ISO 9000 series consists of five standards, ISO 9000 through ISO 9004, that provide guidelines for quality management and quality assurance. In the US, the American Society for Quality Control (ASQC) and the American National Standards Institute (ANSI) "Q90 series" (ANSI/ASQC Q90 - Q94) are now identical to the ISO 9000 - 9004 standards. Ultimately, the US Food and Drug Administration (FDA) is expected to align Good Manufacturing Practices (GMP) and Good Laboratory Practices (GLP) with ISO 9001 (Reference 1). It should be noted that in the US, the ISO 9000 program is not at this time a governmental standards certification program, but rather a program that is administered in the US by ASQC and ANSI (Reference 2).

John Oaklynd, a noted expert in the field of total quality management (TQM), has defined a quality system as an assembly of the components of organizational structure, assigned responsibilities, procedures, processes, and resources for implementing quality management. The ISO 9000 series of standards incorporates these major elements, 20 in all (covered in detail in ISO 9001), which provide the framework and define the quality assurance system for any organization (References 2, 3).

While the ISO 9000 series may provide the basics for a sound quality assurance system, it has its limitations. It has been said that the ISO credo "Say what you do and then do what you say" is in itself a limiting statement, describing more of philosophical approach rather than actual, well-defined QA/QC protocols. ISO 9000 standards are not so much concerned about the methods used in a quality assurance system, but more concerned about how the system works. For example, a supplier is not told what methods to use and so is free to use any methods which could be outdated, inefficient and/or redundant (Reference 1). In any case, the global significance of ISO 9000 principles cannot be understated and any organization wanting to actively compete in the international marketplace sooner or later will have to consider the impact of ISO 9000 on its operations.

When applied to a QA/QC program within a testing laboratory environment, ISO regulations do not address some of the major aspects of that domain, for example, there are no elements that address the quality of science - methods used and accuracy of results. Although laboratory procedures must be documented if ISO 9000 standards are to be met, they are not scrutinized or evaluated. Although the ISO standards are voluminous, their attempt at being "universal" does not cover the concept of acceptable performance for a modern laboratory.

Important decisions, many medical-, health- and food-related, are based on laboratory results. Therefore it is of great importance to have some indication of the quality and reliability of the results. The laboratory and the analyst must use appropriate methods and reference materials to be confident of the analytical result. While ISO 9000 may provide general guidelines for a QA/QC program within a laboratory, these guidelines apply to testing in general and in most cases additional guidance is necessary to take into account the type of testing, methods and specific QA/QC techniques. To fill this gap, Eurachem has published a list of guidelines "Accreditation for Chemical Laboratories", which may be used in conjunction with ISO 9000 standards (Reference 4).

Reference materials provide traceability to national measurement standards. Their use provides a means for verifying method performance and accuracy of results. While the use of reference materials may not be specifically mentioned in ISO 9000 standards, their use in US certification schemes (e.g., State and Environmental Protection Agency (EPA)) is mandated. Because ISO 9000 standards are updated on a routine basis (at least once every five years), the use of reference standards could be incorporated into future revisions of the ISO 9000 standards. It is interesting to note that ISO 9000 principles can be used to define the quality system under which reference materials are produced, but the ISO 9000 standards themselves do not specify the use of reference materials as part of QA/QC laboratory practice.

What argument can then be made for the use of reference materials in a laboratory QA/QC scheme that may be based on ISO 9000 principles?

As stated in the section "Reference Materials, Reference Values of Statistical Tests in the Quality Assurance Schemes of Analytical Laboratories", in the book "Quality Assurance for Analytical Laboratories" (Reference 5), the estimate of the mean (x) in a determination can be tested by use of certified reference materials (CRM's). CRM's for chemical analysis are characterized by knowing concentration values, and therefore, they can be used for monitoring the mean value (x) of an analytical process or for calibrating an instrument. Such materials could also prove useful in attempting to compare and assess laboratory certification schemes.

The matrix of the CRM should approximate or be identical to the matrix of the sample being analyzed. From the point of view of the QA/QC scheme of the laboratory, the parameters x, S_r and S_R are necessary for proficiency testing and the acceptance of S_r is of fundamental importance; however, for the acceptance of these parameters a reference value is needed.

Because of matrix requirements, an actual certified reference material with specific matrix requirements may not be available. In such cases, samples either that have been analyzed "in-house" or that have been analyzed in laboratory proficiency programs may be used as "secondary reference materials."

While CRM's would be preferred, they may be available with the appropriate analyte(s) and/or matrix. As an alternative, "in-house" or "secondary reference samples" may be available. Regarding the latter, the National Institute of Standard Testing (NIST) in the US and the Laboratory of the Government Chemist (LGC) in the UK currently maintain lists; there are no doubt other agencies.

The AOCS, as a spin-off of the Smalley Laboratory Proficiency Program, has available a variety of secondary reference materials. There is currently a joint effort in progress between the AOCS Smalley Program and the AOAC Division on Reference Materials to develop a trans fatty acid (TFA) secondary reference material.

References

1. Cereal Foods World 38(11): 866-867 (1993).
2. NIST, "ISO 9000". (April, 1993).
3. ASQC Report, "RAB Introduces Quality Systems Auditor Certification Program". (1994).
4. Eurachem Newsletter No. 6, May (1994).
5. "Quality Assurance for Analytical Laboratories", M. Parkany, Editor, International Organization for Standardization, Geneva, Switzerland. Published by Royal Society of Chemistry, Cambridge, UK. 1993. pp. 109 - 119.

A Proposal for Statistical Evaluation of Data Resulting from a Proficiency Test

L. Alvarez, L. E. Narváez, A. Pérez, and J. F. Robles

NATIONAL CENTER FOR METROLOGY, CENAM, APTDO. POSTAL 1-100, QUERÉTARO, MEXICO 76000

1 SUMMARY

An attempt is made to compare the evaluation of a proficiency test applying z-score versus the mean squared error.

Since it was the first exercise of an interlaboratory study, the problem using z-score was that there was not a good estimate of the standard deviation, and some difficulty to establish criteria even though the elimination of outlier laboratories by Cochran and Grubbs´ tests was applied.

Using the mean squared error and a certified value, in which both bias and standard deviation on the actual data set reported by the laboratories are considered, and there is no need to look for outlier laboratories. The criteria to evaluate the laboratories would be the target value for bias and standard deviation.

2 INTRODUCTION

Due to the necessity to control the contaminants in waste water for the region of Mexico City and the State of Mexico, the regulatory agencies of the local governments together created a network of testing laboratories capable to do the analysis.

As part of the accreditation procedure of the laboratories, an interlaboratory test comparison was organized. CENAM did prepare and certify reference materials to be used as the certified solutions, distributed among the laboratories registered to participate. The parameters to evaluate were restricted to Cu, As, Cd, Cr and Pb in HNO_3-2% v/v, Hg in HNO_3-2% v/v+K_2CrO_7-3mg/l, CN^- in NaOH-0.5%w/v, and Cr^{+6} in HNO_3-2% v/v.

3 SCOPE

This document is to be addressed as a reference method to be considered in the statistical procedure for the analysis of the results in proficiency testing, either for a single or several parameters.

4 DEFINITIONS

The following definitions are adopted in this paper.

4.1 Proficiency Testing Scheme

Method of checking laboratory testing performance by means of interlaboratory tests.

4.2 Testing Laboratory

A laboratory that measures, examines, tests, calibrates, or otherwise determines the characteristics of performance of materials or products.

4.3 Certified Reference Material (CRM)

A reference material, accompanied by a certificate, one or more of whose property values are expressed, and for which each certified value is accompanied by an uncertainty at a stated level of confidence.

4.4 Assigned Value

The value to be used as the true value by the proficiency testing coordinator in the statistical treatment of results and the best available estimate of the true value of the analyte in the matrix.

4.5 Bias

The difference between the expectation of the test results and an accepted reference value.

Note: Bias is a systematic error as contrasted to random error. One or more systematic error components may be contributing to the bias. A larger difference from the accepted referencevalue is reflected by a larger bias value

4.6 Standard Deviation (of a random variable or of a probability distribution)

[ISO 3534-1, 1.23]. The positive square root of the variance.

4.7 Mean Squared Error

An overall measure of the size of the measurement error, which is defined as

$$MSE=E[(X-x_0)^2].$$

The mean squared error can be decomposed into contributions from the bias (β) and the variance (σ). A perfect measurement would have β=0 and σ^2=0.

5 CERTIFICATION OF REFERENCE MATERIALS AS THE TEST MATERIAL

For the reference solutions prepared by CENAM, the assigned values as well as the associated uncertainties were established by CENAM taking into account the experiences obtained in NIST, High Purity Standards and Hughes Aircraft in the USA.

6 PARTICIPANT LABORATORIES

The total number of laboratories participating in the interlaboratory study was 41, from industry, government and schools.

7 RESULTS USING Z-SCORE

For the sake of illustration only the results of Cu and Cr^{+6} are presented.

7.1 Z-Score

Using the assigned value as the true value, X, the estimate of the bias is defined as:

$$\text{Bias} = x - X$$

where x is the mean value reported by each laboratory.

Since most of proficiency testing schemes proceed by comparing the bias estimate with a target value for standard deviation that forms the criterion of performance, the z-score approach, by evaluating the z-score, defined as

$$z = \frac{(x - X)}{\sigma}$$

is adopted in the laboratory performance evaluation. Instead of σ, the uncertainty associated to the certified value was used, giving more weight to the Type B uncertainty. Since it was the first exercise, there was not previous information to have a good estimate of σ.

7.2 Outliers Elimination

After the z-score was obtained, it was decided to eliminate outlier laboratories by Cochran and Grubbs′ tests. Based on the results, it was found that some labs were eliminated by Cochran′s test, some were eliminated by Grubbs′ test, and some were eliminated by both. So only the laboratories who passed both tests are considered eligible to obtain z-score evaluation. The results are presented in Figure 1 and Figure 2 for Cu and Cr^{+6} respectively.

7.3 Criteria for classification

Classification is not the primary aim of proficiency tests. However, for the purpose of laboratory accreditation, the statistical fundament of the quantiles associated to the distribution of the statistic z-score were applied. In a well-behaved analytical system z-score would be expected to fall outside the range $-2<z<2$ in about 5% of the instances, and outside the range $-3<z<3$ only in about 0.3%. The final classification of the z-scores would be as the following:

$|z| \leq 2$, Satisfactory

$2<|z|<3$, Questionable

|z|≥3, Unsatisfactory

The z-score used as the criteria of classification should be treated with care due to the assumptions on: (1) that the appropriate values of X and **σ** have been used, and (2) that the underlying distribution of analytical errors is normal, without considering outliers.

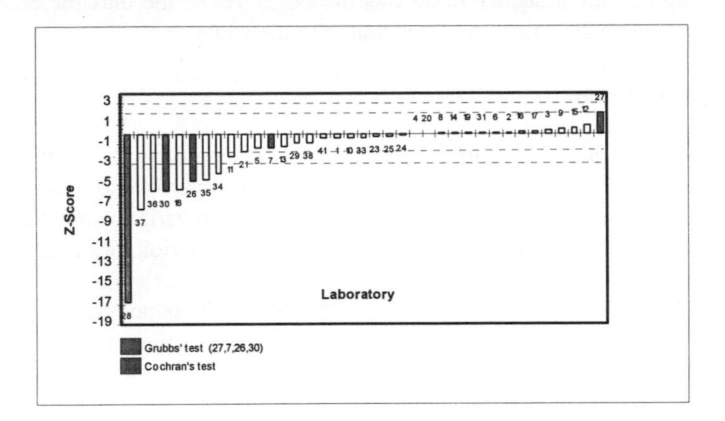

Figure 1 *Z-score for Cu*

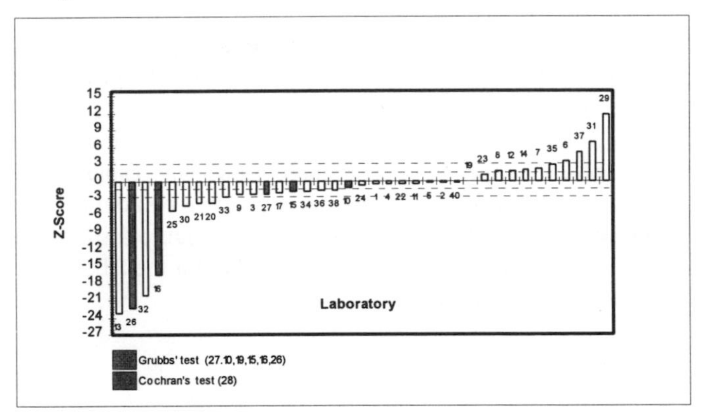

Figure 2 *Z-score for* Cr^{+6}

Following is the classification recommended for consideration:

Cu

|z| ≤2	1 2 3 4 5 6 7 8 9 10 12 13 14 15 16 17 19 20 21 23 24 25 29 31 33 38 41
2<|z|<3	11 27
|z|≥3	18 26 28 30 34 35 36 37

Cr^{+6}

|z| ≤2	1 2 4 5 10 11 12 14 15 17 19 22 23 24 34 36 38 40
2<|z|<3	3 7 9 27 33 35
|z|≥3	6 13 16 20 21 25 26 29 30 31 32 37

8 RESULTS USING MEAN SQUARED ERROR

8.1 Bias and Standard Deviation

Considering that the assigned value was the target value, the bias for each lab was calculated, the standard deviation and the estimate of the variance.

8.2 Mean Squared Error

It is important in doing a measurement to have some idea about the size of the measurement error, and the mean squared error is a very appropriate statistic for this purpose, because it considers the two most important sources of variation together. The bias gives an estimate of the systematic error and the standard deviation, or variance, gives an estimate of the random errors.

Both the MSE and the uncertainty are calculated based on the same principle, so they can be compared and interpretated the same way. Also, the bias and the standard deviation are statistically independent. Having the bias from a laboratory, it can not be said anything about its standard deviation and vice versa.

8.3 Criteria

The goal of the interlaboratory test was to recognize laboratory capability. The criteria proposed here is to use the information of the uncertainty assigned to the certified sample that includes an allowance for the systematic error of the laboratories.

For the laboratories within the uncertainty they are accepted as proficient labs. Otherwise, the laboratories should be encouraged to improve their proficiency. Table 1 shows the results for Cu and Cr^{+6} in which the acceptable laboratories are shown by bold numbers.

In Figure 3 can be seen graphically the MSE for Cu. Figure 4 shows the MSE for 4 elements. In this plot it is clear what was the major problem for the laboratories in terms of bias (accuracy) and standard deviation (precision).

Table 1 *Results for Laboratories accepted as proficient labs to perform analysis of Cu and Cr^{+6} in waste water*

MSE Cu	*Lab*	*MSE Cr^{+6}*	*Lab*
0.0060	**20**	**0.005**	**40**
0.0066	**19**	**0.0058**	**11**
0.0136	**16**	**0.0076**	**22**
0.0154	**6**	**0.0086**	**4**
0.0232	**23**	**0.0089**	**5**
0.0246	**2**	**0.0116**	**24**
0.0259	**31**	**0.0147**	**1**
0.0302	**3**	**0.0153**	**2**
0.0333	**8**	0.0233	38
0.0368	**14**	0.0236	23
0.0394	**4**	0.0238	12
0.0478	**10**	0.0251	8
0.0492	**15**	0.0252	36
0.0586	**24**	0.0286	14
0.0618	**41**	0.0301	7
0.0627	**9**	0.0303	3
0.0650	**25**	0.0305	17
0.0677	**33**	0.0339	34
0.0678	**1**	0.0363	33
0.0764	**12**	0.0366	9
0.0899	38	0.0398	35
0.1171	13	0.0460	19
0.1238	29	0.0463	6
0.1328	5	0.0492	15
0.1508	21	0.0494	20
0.1657	7	0.0500	10
0.2065	11	0.0514	21
0.3381	34	0.057	30
0.3958	35	0.0661	27
0.4405	17	0.067	37
0.4816	18	0.0628	25
0.4830	30	0.0900	31
0.4928	36	0.1537	29
0.6402	37	0.216	16
1.4420	28	0.261	32
		0.290	26
		0.301	13

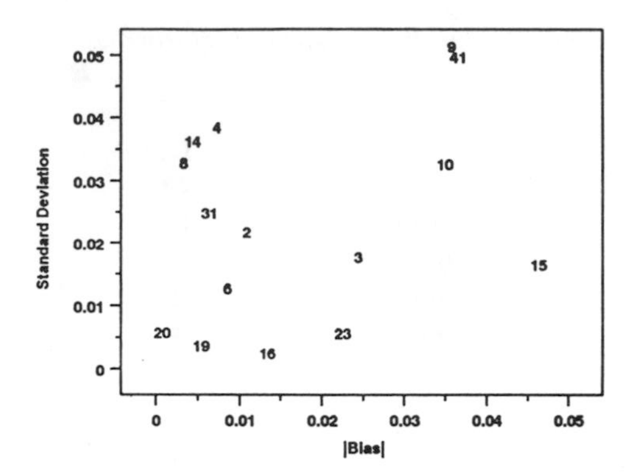

Figure 3 *MSE for Cu*

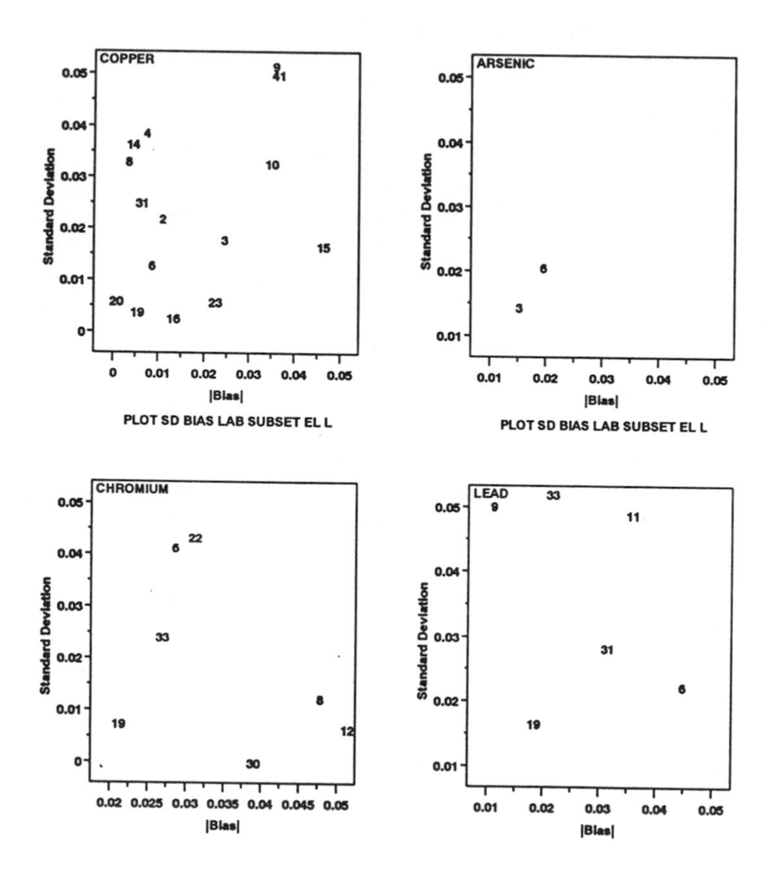

Figure 4 *MSE for Cu, As, Cr, and Pb*

9 A COMBINED SCORE

It might be necessary to assign a combined score for the total of parameters to evaluate. In this case what might be done is to get, element by element, the weighted MSE (using the mean of the MSE as the weight) so the difficulty to analyse an element can be taken into account. Finally the combined score of the MSE is obtained by adding up the individual weighted MSE for each laboratory.

The criteria to say which laboratories are going to be considered as proficient over all the parameters, would be the sum of the individual uncertainties divided by its own weighing factor.

9 CONCLUSION

It is demonstrated that the evaluation of laboratories by MSE is straightforward, without ambiguity.

Acknowledgment

Thanks are due to Dr. M. Levenson and Dr. J. Filliben, from the Statistical Engineering Division at NIST, to members of the metrology of materials laboratory at CENAM, and to the permission of the authorities of Mexico City and the State of Mexico for using the data.

References

1. Thompson and R. Wood, "The International Harmonized Protocol for Proficiency Testing of (Chemical) Analytical Laboratories", 2144 JAOAC International 1993, Vol. 67, p. 926-940.
2. A. Rice John, "Mathematical Statistics and Data Analysis", Wadsworth & Brooks/Cole Advanced Books & Software, Pacific Grove, California.
3. Sutarno, R., "Procedures for Statistical Evaluation of the Data Resulting from International Tests", ISO/REMCO N 285, November 1993.

Teaching and Learning Total Quality Management: Remarks on Aims and Methods Concerning University Studies

László Paksy

METALCONTROL KFT. MISKOLC, VASGYÁRI U 43 H-3540, HUNGARY, AND UNIVERSITY OF MISKOLC, CHAIR FOR QUALITY ASSURANCE, MISKOLC, H-3515, HUNGARY

INTRODUCTION

The process : learning-teaching - learning ... can be assumed as a kind of information process (change of informations) which can be characterized by the following features:

1. Anybody who teaches must firstly learn what he wants to teach;
2. He must continuously renew his knowledge, as the rate of information change is comparable of the development of the given science branch ;
3. The object of any learning study must contain :

   a. basic knowledge and

   b. special one(s).

4. It is unavoidable the development of the intellectual power for the understanding of the future newer knowledge, to become able for improvement . Of course it must be assumed also that both teachers and students are fond of the given science branch.

Summarizing the aims and methods applied till nowadays in field of the very new science

branch : Total Quality Management it can be stated that

a. Beginning from the late 70s till to-day, the teaching was done by the individual

b. The management of quality was perceived as an important challenge to Europe, or as a strategic weapon, with the aim to be among the best; therefore leading companies organized TQM workshops at various levels but at least for almost all peoples of their company;

c. where - the decisive role of the management and

- the leading principle :" satisfaction of the customer "

were successfully emphasized, and therefore in the reports written about these workshops a quite quick breakthrough are generally mentioned.

It is ,however, quite evident that because of the world-wide breakthrough of the aims of the Total Quality Management, the strategic aim of its introduction was modified. Nowadays, the use of TQM methods and systems does not ensure automatically the best positions on the market, but it enables the competitiveness at all.

Considering the fact that TQM became world-wide applied from the late seventies, these aims and methods of teaching and learning (firms-centered educational infrastructures) of the eighties and early nineties seem to be appropriate.

In the future, however, the application of TQM will be such a general demand that its teaching,as an initial training complementing the following firm training will be necessary. In the following , the possible aims and methods of an university education will be discussed.

POSSIBILITIES OF THE UNIVERSITY TQM STUDIES

Considering the fundamental aim of the TQM - " the satisfaction of the customer ", or generalized [1] : as a
" Management philosophy and company practices that aim to harness the human and material resources of an organization in the most effective way to achieve the objectives of the organization " (this is which includes the above general aim), it is impossible to give comprehensive knowledge in this topic in the university.
Apart from the manyfoldness of product/ customer's requirements, the " totality" of the quality management alone is - uncoupled to any product - an undetermined definition. It is quite well-marked the contrast between the attributes :"total" and e.g. "basic " knowledge which can be assumed as a possible learning study in the university.
Despite of these difficulties, there are such common principles in all TQM applications, or methods, techniques that can be successfully taught in the universities.
They can be divided into two main groups :

- TQM basic and
- TQM know-how.

TQM BASIC

The main components of a TQM BASIC learning study can be :

1. Fundamental aim ,
2. Definitions ,
3. TQM structures / organizations ,
4. Related topics from the Civil Right,
5. Psychology ,
6. Statistical methods

To the above mentioned items, the following short comments can be added :

- ad "fundamental aim": here the history of the development of the TQM must be also taught , demonstrating the necessity of its application as well as the organic character of this development; the production of wares of good quality is imperative not only from point of view of market success, but from that of energy savings, environmental protection, because of the shortage of various resources (besides ever increasing population on the Earth, etc.). This way,. TQM, as a challenge of our (and probably of the next) century,too, must be placed in a larger relationship showing its coherence with other fundamental problems.

ad :" Definitions ": though it is evident that the corresponding ISO Guides or the new BS 7850 can be successfully applied , the students must be warned against the dogmatic use of these definitions, e.g. the management leadership cannot be interpreted without the responsibilities for the whole quality improvement process in a given situation .

ad " TQM structures / organizations" : this part of the TQM basic has a descriptive character, it must be emphasized that there is no " panacea" for the best structure or organization, besides some examples of educational value (eventually:from the literature) some living examples (possibly: existing TQM systems , they can be visited) may be mentioned. The students must learn that they will be obliged to create new versions for TQM structures , if necessary, and not copy other , not adequate ones.

ad : " Related topics from Civil Right ": it is essential to know the corresponding laws in the given country or in the international trade, e.g. those that refer to the product responsibility etc. This part of the learning studies must be accomplished by a " double" specialist who has been working both as lawyer and manager knowing well both area.

ad: " Psychology" : the fundamentals must be taught by a specialist emphasizing and giving full details about the relations between manufacturer/vendor/customer or the psichology of these relations, respectively.

ad " Statistical methods" : if the prescribed or recommended statistical methods were included in the previous studies already, it is advised to practise them; for their successful application.

TQM KNOW-HOW

The students must be acquainted with the various techniques and tools used -among others- in TQM systems,too. Such are :

- various forms of communications;
- computers/ computer-networks ;
- various tools and techniques for product qualifications, assessment of the performance (e.g. control charts, flow-diagrams , Pareto-diagram etc..,)
- measurement technique

In connection with these learning studies the following must be noted here :

ad " various forms of communications ": the building up of personal contacts with the customers, publicity, oral or poster presentation of plans, results etc., may be the subject matter ;

ad " computers/ computer networks " : I suppose this topic is such a requirement everywhere that it needs no further comments ;

ad " various tools and techniques for product qualifications, the assessment of the performance": the items mentioned above must be practised by means of real examples.

ad " measurement techniques": regarding the variety of measurement methods, as well as tasks , the aim of this learning study is not to be acquainted with all (many) kinds of measurement techniques, however, to know in a given situation what to choose, what are the conditions of a reliable measurement at all, to perceive the existence of the measurement error, etc. Roughly summarizing our learning study of this kind contains the following items :

A. History of the metrology, measuring system: SI, conversion to other measuring systems;

B. Measurement , and measurement methods of the 7 ground quantities of SI (time, length, mass, temperature, electric ground quantities, mass-quantity, luminous flux);

C. Characteristic parameters of the instruments, their calibration ; norms for the measurements (domestic and international)

D. Traceablity and uncertainty- definitions , calculations .

E. Special measurements in the given science branch : here :

- for metallurgical engineers students:

  measurement of the mechanical properties of the steel (strength testing, impact test, impact-endurance test, creep testing, hardness test, Jominy -test, dilato-metric investigations);

F. Accreditation, validation of the measurement methods.

TQM IN OTHER LEARNING STUDIES

It is evident that some elements of the above mentioned learning studies can/ must be included in other learning studies.

Some examples :

- by teaching various production technologies , the role and importance of TQM can be included ,too;
- statistical methods can be taught in frame of the mathematical studies .

Apart from these evident possibilities, however, the spirit of TQM can be interpreted in many learning studies.

As the " total" quality management can be explained only in relation of a given production system , similarly, the "theoretical" studies must be complemented by practical exercises . These can be formed according to the circumstances,possibilities.

ABSTRACT

Remarks are summarized regarding complementary learning studies in Universities, re: TQM . The learning study : "TQM" can consist of two main parts: TQM basic, and TQM know-how, and the possible content of this study is discussed in detail .

LITERATURE

[1] Total Quality Management, Part l. Guide to management principles ; Part 2.Guide to quality improvement methods. British Standard BS 7850. 1992

Quality in the Analytical Laboratory: from 'I' to 'We'

W. G. de Ruig[1] and H. van der Voet[1,2]

[1] DLO – STATE INSTITUTE FOR QUALITY CONTROL OF AGRICULTURAL PRODUCTS (RIKILT-DLO) PO BOX 230, 6700 AE WAGENINGEN, NETHERLANDS

[2] DLO – AGRICULTURAL MATHEMATICS GROUP (GLW-DLO) PO BOX 100, 6700 AC WAGENINGEN, NETHERLANDS

1 FROM ISOLATION TO INTEGRATION

From 'I' to 'we', that means taking procedures of chemical measurement out of the individual sphere to an integrated approach.

In the past, say until the 1960s, each analytical laboratory worked alone. 'I am right, because I did it and I did it well.' And I was striving for success — or failure. Conceding that my results were wrong and those of another laboratory were right amounted to resignation and admission of incompetency. This attitude gave huge problems in disputes about international trade and forensic investigations. Each party claimed it was right and was equally eminent (Figure 1). To overcome the impasse, collaborative studies since the 1960s have been widely used to validate standard methods. At first, the results were disconcerting: it was not expected that the results of intercomparisons would be so bad. Within-laboratory variability had been assessed

Figure 1 *Who is right?* (by courtesy of Eurachem)

too favourably, for instance by rejecting 'bad' results and taking only the 'best' results.

These studies have improved understanding of the quality of the analysis.

Firstly, competent laboratories collaborated not to prove they were right but to test the reliability of a method, accepting each other for their worth. Gradually an open atmosphere was created: if there is a wide variability in results, it was not discredit that one person's results differed from another. The uncertainty of methods became clear. Scientists learned that 'errors' were not 'mistakes'.

Secondly, such an open atmosphere gave the opportunity for a cooperative attitude: workers looked together for the sources of variability and tested together how to minimize it. In this way, methods were improved substantially.

A statistical foundation was essential for this process. Analytical chemists tend to see one measurement as an isolated action, and a series of multiple measurements as a repetition of this action. Statisticians consider the distribution from which the analytical result is just one. The cooperation of statisticians and analytical chemists was an important step in the way from isolation to integration.

2 LABORATORY ACTIVITIES AS A PRODUCTION PROCESS

We have to recognize that an analytical chemical laboratory is like any other production unit. Our products are analytical data and there are customers who pay for them. Like any other product, e.g. cars, our products too must be manufactured in a clearly defined and stable production process subject to stringent procedures of a quality assurance system. Policy makers in industry and government base their strategies on analytical results. The consequences can be immense.

Thus we have to transfer the ideas of general quality control and assurance to analytical chemistry. One of the main considerations that will change many of the existing operating procedures is that analytical data is produced in a *process*, not in an isolated action. The traditional order in operating procedures was (1) prepare equipment and chemicals; (2) analyse; (3) calculate results. This should be replaced by something like: maintain the analytical process in a state of control, which is continually checked by quality-assurance samples; analyze samples from customers together with these quality-assurance samples under a specified schedule.

3 USEFUL CONCEPTS FOR INSPECTION AND FORENSIC PROCEDURES

Let there be a legally permitted maximum value, M, of a substance (here taken as a concentration, c). Zero tolerance corresponds to $M = 0$. (For minimum values, the reasoning is equivalent.) To test whether or not the sample meets the requirement, we can use the null hypothesis, H_0: $c \leq M$. The alternative hypothesis, H_1, is $c > M$. For the quality of the procedure, only two criteria are relevant:

- probability of a false positive result (α) for samples with $c = M$. This is the *maximum* probability of a false positive result for samples with $c \leq M$.
- probability of false negative results (β). This is a function of c. Only for higher concentrations is there a combination of low α and low β.

To keep α small, the decision for disapproval has to be taken at a concentration higher

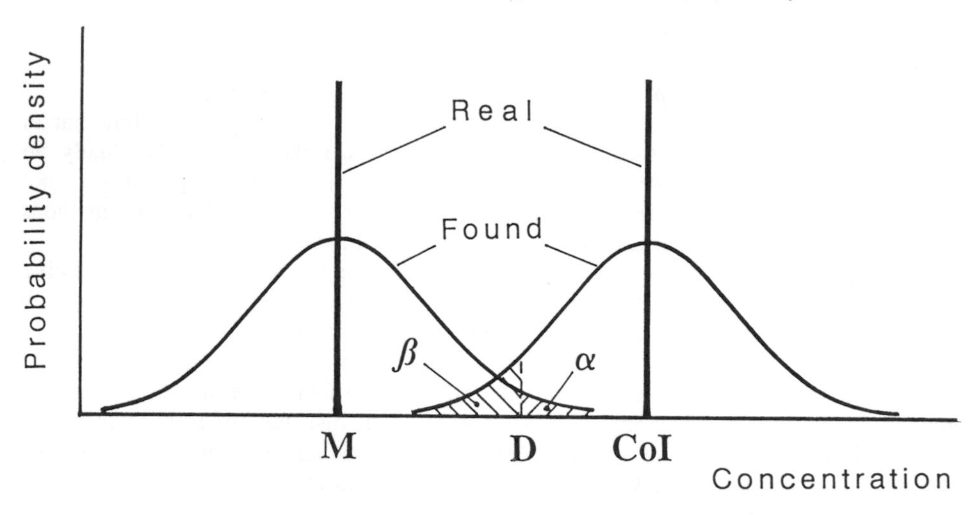

Figure 2 *Quality characteristics of an inspection procedure*

than *M*: the *decision limit*, *D*; that is the value used by the inspecting laboratory, to decide on positive or negative results (Figure 2). We introduce the *capability of inspection*, or the *minimum detectable inadmissible concentration*, *CoI*. That is the concentration at which the probability β of wrongly accepting the inspected lot is acceptably small. Below *CoI* the laboratory is not 'capable' of performing the inspection adequately, because β is too high. We have to specify what is an 'acceptably small' value for β. Here the scope of 'we' widens: an effective inspection control system cannot be built by analytical chemists alone. It has to be designed and implemented together with legislators, policy makers and statisticians.

It is the duty of legislators and policy makers not only to specify the tolerance level *M* but also the capability of inspection, *CoI*, together with the α and β they require. — And also they must provide the funds. Higher requirements cause more effort by the laboratory and higher costs. The laboratory's duty is to prove that its measurement method is suitable for the inspection task, i.e. that its capability of inspection does not exceed the specified value. To this end, each inspection laboratory has to determine its own quality characteristics, and to prove that these are equal to or better than the minimum requirements. Statisticians have to help both decision makers and analytical chemists to enable a sound procedure, e.g. by offering an appropriate model for the calculating and evaluation of results.

4 HARMONIZATION

So more and more laboratories were working together to establish the quality of what they were doing. But then the question arose: In which way 'we' must carry out our performance tests? There was a need for reference materials and reference methods. A plethora arose of international standards, guidelines and protocols to test and control the quality of measurement data in laboratories. Some of the main ones are listed below.

4.1 Harmonized protocols

A series of harmonized protocols have evolved and continue to be evolved under the auspices of the International Union of Pure and Applied Chemistry (IUPAC), the International Organization for Standardization (ISO) and AOAC International. They include protocols for design, conduct and interpretation of collaborative studies,[1] for adoption of standardized analytical methods and for the presentation of their performance characteristics,[2] for proficiency testing of chemical laboratories[3] and for total quality management.[4]

4.2 Quality assurance and accreditation

Various international standards and guides deal with quality assurance in analytical chemical laboratories.

The most widely accepted international quality guide for testing laboratories is ISO/IEC Guide 25 on General requirements for the competence of calibration of testing laboratories.[5] The International Standard series ISO 9000[6] deals with quality assurance in relation to a product, production process or service and is particularly intended for industrial companies. It focuses on an efficient and reliable management of affairs and is widely accepted in industry on a voluntary base. Laboratory evaluation is not a primary aim.

The European Standard series EN 45 000[7], largely based on ISO Guide 25, is aimed at quality assurance of information produced by laboratories concerned with testing, inspection and calibration. For the requirements of a laboratory quality assurance scheme, EN 45 000 and ISO 9000 have the same intention. Additionally, EN 45 000 also considers the competence of the laboratory. Laboratory accreditation under EN 45 000 guarantees that the laboratory can meet the procedural requirements for specified measurements. Generally it is voluntary but in some EU countries it is obligatory, even legally binding.

The target group for good laboratory practice (GLP) is laboratories responsible for non-clinical safety studies, aimed at submission of dossiers for registration by national licensing bodies. In the 1970s, the US Food and Drug Administration (FDA) evaluated some dossiers; too much of the research and the ensuing reports proved to be so careless that the results were unreliable. Several instances of fraudulent research were also discovered. These incidents led the FDA to establish general rules[8], which passed into US law in 1979.[9] As a result, all safety research, i.e. toxicological studies and research into licensing of animal and human drugs, additives and cosmetics had to comply with these rules of GLP. Likewise the US Environmental Protection Agency (EPA) has formulated GLP regulations intended for laboratories involved in environmental research.[10,11]

A Code of Good Laboratory Practice was also drawn up by the Organization for Economic Cooperation and Development (OECD),[12,13] which modified the FDA guidelines to match the administrative practices of the various member states. They are now being used in many industrialized countries. The European Union (EU) implemented the OECD principles as a European standard.[14]

There is some overlap between the criteria for accreditation and GLP. Eurachem and Eurolab (organizations of laboratories) are developing a harmonized guidance document for simultaneous implementation in a single organization.

The EU has tended to abandon developing methods by itself. Instead EN standards will be used in EU legislation. Unlike ISO standards, these EN standards are mandatory. In its turn, CEN is adopting other standards, e.g. of ISO, where appropriate.

National bodies are concerned with laboratory accreditation in many countries all over the world. Two European organizations concerned with accreditation merged in 1994, WELAC (Western European Laboratory Accreditation Cooperation) and WECC (Western European Calibration Cooperation), form a single organization EAL (European Cooperation for Accreditation of Laboratories).

5 SHERLOCK HOLMES

As indicated, we can work together and statistics can be helpful in the laboratory. But the approach works only for series of samples. What about one unique investigation? The hair, found by Sherlock Holmes at the place of the crime? Statistics then fail. 'I', the individual, is responsible for the analysis. Evidence for the correctness of the result must be based upon the expertise of the investigator or on analogy with similar results, not on statistical validation. Sherlock Holmes is a typical 'I' person.

Therefore for the 'we' approach, the definition of the population of samples the laboratory can analyse under validated procedures is all-important. It was formerly inadequately recognized how important it was to decide beforehand on the scope and limits of a method.

To continue the analogy of the hair, if it arrives in your laboratory only incidentally, it may not be cost-effective to set up and to maintain a validated method in your laboratory itself. Perhaps a specialized laboratory is prepared to analyse samples of this kind, and it is better to contract out analysis for them. Laboratories with no experience in analysing 'hairs' should refrain from such analyses, even if they see no analytical problem. Have no regrets if you cannot do everything: that is another lesson 'we' have learned.

6 CRITERIA

In the 1980s, unequivocal description of methods combined with validation by collaborative studies proved to be no final solution.

Exact formulation of analytical procedures has some drawbacks.

(1) A standardized method, by virtue of its rigidity, cannot accommodate new developments, and is likely to become old-fashioned.
(2) Modern methods, involving highly sophisticated equipment or materials, may operate according to unique procedures not transferable to other laboratories.
(3) Behaviour of reagents and equipment may vary over time, forcing adaptation of the method.
(4) Information present in an analytical result may not be clearly accessible. This is particularly so for evaluation of multivariate data.
(5) The idea that one well described method will give the same results in all laboratories is wishful thinking.

Yet methods have to be validated before they are acceptable. To solve this dilemma, validation of methods was approached by another way, by formulating 'criteria'.[15] Such criteria have been implemented in EU regulations for detecting residues.[16,17]

Figure 3 *Butter promotion in strip form*

At the beginning of this century, the Dutch painter Willy Sluyters (1873-1949) made a series of crayon drawings on the theme 'Quality control of butter' under commission of the Dutch Government. They were aimed to convince customers in importing countries about the quality of Dutch butter. The drawings were shown at a dairy exhibition in England in 1908 and are now in the conference room of RIKILT-DLO. The texts are from advertising postcards, at that time reproduced from the drawings.

Grocer A: I buy this butter because it bears the Governmental mark, which is a guarantee that it is not adulterated and it contains less than 16 percent of water.
Grocer B: Are you sure about that?
Grocer A: Yes. But if you are in doubt, we may just as well ask the analyst from my district.

The approach was adopted by the Codex Committee on Methods of Analysis and Sampling.[18]

There is another reason why there will be less need for standards on analytical methods. To become accredited, a laboratory has to implement a quality handbook describing all procedures within the laboratory. For all methods of analysis, standard operating procedures (SOPs) are required. Like standards, they describe precisely and unambiguously the method, but unlike the latter, SOPs are flexible and can easily be adapted to new developments. The quality requirements by the accrediting authority are described in general terms only, for instance precision (repeatability and reproducibility), trueness, specificity, applicability, and limit of detection.

7 INTEGRATED APPROACH

The idea of an integrated approach instead of an isolated action is not completely new. At the beginning of the 20th Century, there was an interesting discussion[19] between

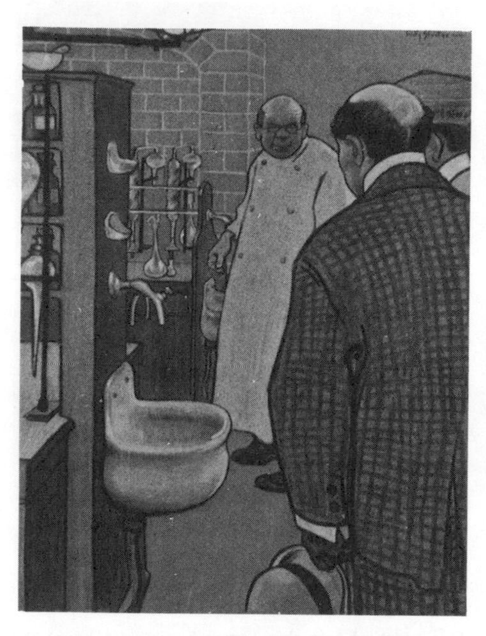

Grocer B: Are you convinced by your analysis that the butter bearing the mark of the Netherlands Government is absolutely pure?
Analyst: Yes, I am, because the results of my analysis of a sample of tested butter always agree with the composition such butter must have according to the information given on my request by the officials in the Netherlands.
Let us make a test with the butter we have just seen.

Analyst: From a telegram from Holland I learn the chemical composition which the butter bearing the governmental mark E-Bd 96347 ought to have. My analysis of the sample is in perfect agreement with that and so the butter cannot be adulterated.
Grocers A and B: Really we are now convinced that the Netherlands Government mark gives a full guarantee. In future, we will accept from the Netherlands only butter that bears the Government mark.

Van Sillevoldt, first director of the Dutch Government Dairy Station[20] and his opponent Salomonson.[21] Adulteration of butter was checked by measuring characteristic values: Reichert-Meissl value for volatile fatty acids; content of solid free fatty acids; refractive index; iodine value. Salomonson declared that 'the chemist cannot generally prove that a foreign fat, not originating from milk, is present in a suspect butter of unknown origin.' Van Sillevoldt argued that one characteristic quantity may be insufficient, but that a combination of characteristic quantities could establish that the butter was unadulterated. Translated in modern chemometric language: by multivariate data analysis. Moreover, not only the butter as such was investigated, but also the places of manufacture at the butter factories, and the composition of the butter was compared with that of the cream (Figure 3). Nowadays this approach is called 'integral chain control'. It is a topical approach in food testing. This is a further broadening of the 'we' circle: origin of the product, treatment, sampling, analysis and data evaluation are drawn together in one concept.

8 CONCLUSIONS

The 'we' approach has entered our laboratories. It helps us in introducing and applying standardized procedures.

Our aim is to produce analytical data of constant, specified quality. We have to implement what we mean with 'quality' and to demonstrate how we can measure and maintain it. 'I' alone cannot provide a solid quality assurance system. As we saw above, 'we' together have passed some milestones:

The more open 'we' atmosphere brings opportunities to relativize and optimize. Working 'under GLP' or as an 'accredited laboratory' provides some guarantee for the quality of our product. The procedures can be traced beforehand or afterwards. Fraud has become more difficult, but still cannot be excluded. We can offer our clients a product of agreed quality — and price.

Have we attained the ideal situation yet? Certainly not! We are only just starting. Implementing the rules of a quality-assured production process into the laboratory is a huge effort and investment. At first, only the costs and not the benefits are to be seen. There is lack of understanding between various disciplines. In contrast to the 'we' approach, competition between laboratories and organizations tends towards isolation instead of cooperation.

Yet we have to be aware of the necessity of the 'we' approach. We have to know what 'we' are doing. We can then convince our authorities that it is beneficial for them to prevent wrong decisions, which would waste much more than 'we' cost!

References

1. W. Horwitz (ed.), *Pure and Applied Chem.*, 1988, **60**, 855.
2. W.D. Pocklington (ed.), *Pure and Applied Chem.*, 1990, **62**, 149.
3. M. Thompson & R. Wood (ed.), *Pure and Applied Chem.*, 1993, **65**, 2123.
4. Under investigation.
5. General requirements for the competence of calibration and testing laboratories. ISO/IEC Guide 25. International Organization for Standardization, Geneva, 1990.
6. Quality management and quality assurance standards - Guidelines for selection and use. ISO 9000:1987 = EN 29000.
   Quality systems - Model for quality assurance in design/development, production, installation and servicing. ISO 9001:1987 = EN 29001.
   Quality systems - Model for quality assurance in production and installation. ISO 9002:1988 = EN 29002.
   Quality systems - Model for quality assurance in final inspection and test. ISO 9003:1987 = EN 29003.
   Quality management and quality system elements - Guidelines. ISO 9004:1985 = EN 29004.
   International Organization for Standardization (ISO), Geneva - Organisation Commune Européenne de Normalisation (CEN/CENELEC), Brussels, 1985-1988.
7. General criteria for the operation of testing laboratories. EN 45001, EN 45002. General criteria for laboratory accreditation bodies. EN 45003.
   Organisation Commune Européenne de Normalisation (CEN/CENELEC), Brussels, 1989.

8. Good Laboratory Practice Regulations, Final Rule. US Food and Drug Administration, Federal Register 52, 33782, 4 September 1987 (update).
9. Good Laboratory Practice in Non-clinical Laboratory Studies. US Food and Drug Administration, Title 21, Code of Federal Regulations, Part 58 (21 CFR 58), 1979.
10. Good Laboratory Practice Standards. Federal Insecticide, Fungicide and Rodenticide Act (FIFRA), US Environmental Protection Agency, Title 40, Code of Federal Regulations, Part 160 (40 CFR 160), 1989.
11. Good Laboratory Practice Standards. Toxic Substance Control Act (TSCA), US Environmental Protection Agency, Title 40, Code of Federal Regulations, Part 792 (40 CFR 792), 1987.
12. GLP in testing chemicals. Final report of the OECD Expert Group on Good Laboratory Practice, Paris, 1982.
13. OECD Principles of Good Laboratory Practice. Decision of the Council of 12 May 1981 concerning the mutual acceptance of data in the assessment of chemicals, C(81)30. Organization for Economic Cooperation and Development, Paris, 1981.
14. M. Thompson & R. Wood (ed.), *Pure and Applied Chem.*, 1993, **65**, 2123.
15. W.G. de Ruig, R. Stephany and G. Dijkstra, *J. Assoc. Off. Anal. Chem.*, 1989, **72**, 487.
16. Commission Decision of 14 April 1993 laying down the methods to be used for detecting residues of substances having a hormonal or thyrostatic action (93/256/EEC). *Off. J.*, 1993, **L 118**, 64.
17. Commission Decision of 15 April 1993 laying down the reference methods for detecting residues (93/257/EEC), *Off. J.*, 1993, **L 118**, 75.1722.
18. Report of the 19th Session of the Codex Committee on Methods of analysis and Sampling, Budapest, 21-25 March 1994. Codex Alimentarius Commission, ALINORM 95/23, Rome, 1995.
19. E.A. de Ruig, *Chemisch Magazine*, 1994, 389.
20. H.E.Th. van Sillevoldt, *Chemisch Weekblad*, 1905, **2**, 505.
21. H.W. Salomonson, *Chemisch Weekblad*, 1905, **2**, 481.

Harmonization of Quality Assurance of Analytical Laboratories of MERCOSUR

Alfredo M. Montes Niño

XENOBIOTICOS SRL, BOLIVIA 5826/28, BUENOS AIRES, 1419 ARGENTINA

1 INTRODUCTION

The 1st. of January of 1995 it has started a Customs Union among the four countries that have decided to create MERCOSUR (Mercado Común del Sur-Common Market of the South). (1)

During the previous time a number of objectives have been achieved, several activities still have not been completed and many actions relationed with quality control, acreditation, certification and quality assurance have taken place.

These activities have been considered as a permanent working issue that justified an specific commission of MERCOSUR.

CREATION OF THE COMMISION OF INDUSTRIAL QUALITY

Taking on account the necessity of facilitating the free circulation of goods and services among the member countries and therefore eliminating all types of non tariff

(1) The regulatory structure of Mercosur begun to shape through significants treaties and agreements: The Integration, Cooperation and Development Treated, November 21, 1988; The Treaty of Buenos Aires, July 6, 1990, setting December 31, 1994 as the date on which a common market between Argentina and Brazil should come into being; the Treaty of Asunción of March 26, 1991, established the creation of Mercosur common market by Argentina, Brazil, Paraguay and Uruguay on December 31, 1994, this treaty was ratified by all four Congresses in November 1991.

barriers to trade, it was created in 1991 the Commission of Industrial Quality whose main objective was the harmonization of standards and procedures relationed with the acreditation, certificationa and testing.

As results of the work of this Commission we can mention among others, the common adoption of international standards, a considerable number of national structures and organizations were mutually recognized, agreements about regulations for acreditation and certifications bodies, and among testing laboratories.

It was considered also the necessity of avoiding the possibility that trade products prices were increased due to the lack of mutual recognition of procedures relating to the quality control evaluation required by the markets or national authorities of the member countries.

One of the most difficult aims is how to promote the balance the different levels of development of the quality systems of each member country. Quality organizations of the member countries will be interconnected.

The acknowdlegement of MERCOSUR's certifications and testing by third countries was also considered also essential to reach competitiveness in the trade with those countries.

Mercosur institutions:

The Mercosur Council: The principal political body, integrated by the presidents, economy and foreign affairs ministers.

The Common Market Group: The executive organ integrated by representatives of the Foreign Ministry, Ministry of the Economy and the Central Bank.

Eleven working sub-committees relating to trade policy, industrial, agricultural and transportation topics, technical standards, etc.

Ministerial meetings (Economy, Education, Justice and Labor); Specialized Meetings (Tourism, Science and Technology and Environment).

The Joint Parlamentary Commission, with representatives of the four parliaments.

It has been proposed to the highest level of decision of Mercosur: the Common Market Group, integrated by the Foreign Affairs Ministers of the Member Countries that is necessary to mantain this Commission after the start of the Custom Union as the subjects for harmonization, acknowledgement and agreement are affected by international, regional and national modifications that take place permanently.

PROPOSAL

1. It was proposed that the Commission of Industrial Quality would continue its activity as a permanent office of MERCOSUR.

2. The main objectives of this office would be:

- Continue with the activities established in the so call "Las Leñas Schedule" (2).

- propose the politics for the negotiation with EU, NAFTA and others regional groups.

- Coordinate the position of Mercosur to be presented in various international forum: ISO, IEC, etc.

- Generate systems for supporting national sctivities relationed with quality.

- Promote international cooperation for development of the above mentioned activities.

ADVANCES OF THE COMMISSION OF INDUSTRIAL QUALITY ESTABLISHED BY LAS LEÑAS SCHEDULE.

(2) The "Las Leñas Schedule" was signed by the presidents of the four countries in June 26th 1992, and established the schedule and subjects of harmonization.

Agreement on harmonization and acknoledgement of the structures of certification and testing.

- Visits for aknowledgement of the national structures relating to quality evaluation:

- Brazil was visited on July 1993, Argentina was visited in March and Uruguay in March 1993.

- In March 1994 a Round Table on Acreditation with the objective of exchanging experiences in auditing laboratories and accreditation systems.

- A survey of mandatory certification proyects in the four members countries.

- Agreements took place among standards organizations of the member countries and the following organisms:

UL - (Underwriter Laboratories)
DQS - (Deutsche gesellchaft zur Zertifizierung von Qualitátmanagementsystem mbH)
SABS - South African Bureau of Standards.
NKA (United Kingdom)
IMQ (Italy)
SACI (People's Republic of China)
CERMET (Italy)
DIN/DGWK
AENOR (Spain)

Activities of permanent technological assistance in certification for the Subgroups of Mercosur and its Commissions.

Two decisions of the Common Market Group were issued in order to promote the use of Quality Management schemes following international standards:

RES. 5/92 - Promotion an Use of the ISO 9000 Standards and ISO/IEC Guides relationed with certification and testing laboratories.

RES. 9/92 - Use in MERCOSUR of the certification systems proposed by the International Standard Organization (ISO).

Agreement about conditions for the mutual acknowledgement of the structures of certification, laboratory acreditation and inspection services (Nov/94).

A decision of the Common Market Group was issued with recommendations for the certification structures in MERCOSUR:

RES. 40/92 - Approval of recommendations about conditions to be accomplished by the structures of certification of products, process and services and quality evaluation in the member countries.

RES. 40/93 - National structures of acreditation.

Basis to elaborate a unified list of certifying companies of the MERCOSUR countries.

It was agreed to elaborate a list of certifying companies of the MERCOSUR countries with certification of their quality systems according with ISO 9000. The list will be issued each six months.

ARGENTINE NATIONAL SYSTEM OF STANDARDS, QUALITY AND CERTIFICATION

As an example of the improvements in quality assurance I will summarize the aspects of this system of recent creation. This is the second country to implement a national system, the first was Brazil.

This National System has three levels (3):

Although it has not yet established, the matters relationed with quality assurance in analytical laboratories will be discussed at Level 2, Committee of Acreditation of Testing Laboratories.

QUALITY ASSURANCE SYSTEMS ACTUALLY HARMONIZED

The harmonization actually reached was the result of international requirements. Laboratories in the area of Atomic Energy have developed quality assurance systems according with the regulations of the International Organization of Atomic Energy.

---

(3) National System of Standards, Quality and Certification

a) Level 1: The National Council for Standards, Quality and Certification, as the highest organ that governs and administrates standarization, quality and voluntary certification, and the Advisor Committee that acts as a consulting organ of the Council.

b) Level 2: The standarization Organism, national institution responsible for the issuing and updating of the standards, and the Acreditation Organism, responsable of: acreditation of the organisms that certify quality systems, products, services and process; acreditation of the testing laboratories and calibration laboratories; and certification of auditors of quality systems.

c) Level 3: Organisms of Certification of quality systems, products, services and process and of the laboratories of testing and calibration.

Total Quality Management and Continuous Improvement in an Integrated Forensic Science Organisation

David N. Gidley and Michael J. Liddy

VICTORIA FORENSIC SCIENCE CENTRE, VICTORIA POLICE, FORENSIC DRIVE, MACLEOD, VICTORIA 3085, AUSTRALIA

1 INTRODUCTION

Much has been written in recent years about quality management, best practice, accountability, high performance work systems, benchmarking, accreditation, etc. It is clear that as the world shrinks in terms of rapidity of information exchange and communications, the influence of change trends is felt sooner.

The international requirement for quality standards in manufacturing and service industries impacts on forensic services. There is an onus to demonstrate to our clients that we are providing quality services from a managed program based on international standards:

- increased technological capability has added a power dimension to forensic science evidence which attracts greater scrutiny because of the impact this evidence has in the courtroom;
- increased scrutiny goes beyond the technology to the individual, the processes and procedures used, and to the laboratory's quality system;
- the laboratory systems therefore must be in compliance with recognised and accepted standards deemed appropriate by the relevant scientific community.

The quality management program must also be subject to audit, periodic review and external accreditation.

Accountability is now a frequently encountered term in both private and public sector organisations. The recent increased emphasis on accountability by governments also reflects the requirements for a quality systems approach. One's accountability must be measured against something and a fully documented quality system provides that basis. The development of the Victoria Forensic Science Centre's (previously State Forensic Science Laboratory, SFSL) continuous improvement program (CIP) is discussed below.

2 DEVELOPMENT

Some five years ago the Victoria Forensic Science Centre (VFSC) embarked on the development and implementation of a CIP (or total quality management, TQM) system in response to an analysis of the current environment, our performance and a perceived requirement to improve our systems as a basis for improving our position for the future.

The program was to cover all aspects of our operations in the field, in the laboratory and in the courts, in accord with relevant international standards such as ISO Guide 25[2]. It would also be in accord with the ISO 9000[1] series covering all activities that effect the quality of the service, the documentation of these activities in an organised manner and in a format that allows everyone to understand and follow these instructions. These standards were recognised as being relevant to forensic science service provision and science services internationally.

Both ISO 9000[1] and ISO Guide 25[2] clearly indicate the rôle of top management and their responsibility in setting and leading the quality systems program for the organisation. This is a key issue as there is a requirement for policy and objectives to be formulated, for responsibilities, authority and resources to be assigned, and for the establishment of systematic and periodic review.

These were important issues in the development of our strategic approach to establishing a comprehensive TQM system. Equally important was maintenance of forensic service delivery without disruption from the huge task of reviewing, re-formatting and re-writing the documentation.

Commitment to the program by senior management is essential and this was evident in VFSC's experience. The program was initiated at Directorate (top management) level and the education process started at that level and from there progressively throughout the organisation. The degree of permeation of the proposal varied across the organisation and this was not unexpected given the broad service mix of the VFSC. Attention was paid to those areas that were slower to accept change. In our experience it proved successful to focus on training programs as a means of demonstrating the merit and value of implementing quality systems. Ownership of the manuals produced was a major step forward in the acceptance of the program. The information/education requirement cannot be overemphasised as a commitment to CIP involves a cultural as well as a procedural change.

To underline the commitment, a quality manager, reporting directly to the Centre's Directorate (and the Director, as necessary) was appointed and the Quality Management Branch (QMB) formed. While this Branch was a focal point for most of the activities of the program, senior management provided an ongoing information focus to reinforce the change of culture as to how the Centre's service was to be delivered under CIP. This was reflected in our implementation program.

3 IMPLEMENTATION

From the outset, the quality management system was to be implemented across the entire Centre including not only the scientific/technical areas, but also the support services and administration. As the Centre provides a wide range of services in biological, chemical, drug and technical disciplines, it was a large task to take on. Because of this a senior manager was allocated to the implementation program full time for twelve months. This was in addition to the already appointed Quality Manager. Our experience was that implementation was a different and separate task to the ongoing management of the program and was therefore handled differently with different people leading the separate tasks.

The strategic framework developed was based on the American Society of Crime Laboratory Directors-Laboratory Accreditation Board program (ASCLD-LAB)[4], which aligns with ISO 9000[1] and ISO Guide 25[2] in requiring forensic services seeking accreditation to not only have a quality system in place, but also to have it appropriately documented. Our program resulted in the production of:

- laboratory manuals of three volumes
- methods manuals and training manuals across 25 Sections
- Division operating procedures manuals
- a Laboratory Safety and First Aid Manual, and
- a Standard Reference Works Manual.

All methods manuals were written in conformity with the Standard[3] for analytical chemistry methods; this necessitated lateral thinking in applying the Standard to procedures for crime scene examination and recording, for example.

As this accreditation program (ASCLD-LAB) was forensic science based, it proved to be readily acceptable and useable by all staff throughout the organisation. The ASCLD-LAB program provided the structure, the format and the time frame for the implementation of the Centre's quality management system.

Throughout the implementation period, staff were provided with information through meetings, workshops, seminars and via their own area's achievements in completing training, methods or procedures manuals. Progress of the program was also plotted on three large logistics boards available for all staff to peruse and the boards were photographed periodically.

Working with the ASCLD-LAB program[4] identified some limitations, particularly in regard to highly integrated forensic services with a major service commitment in crime scene examination. The program covers eight forensic disciplines but these do not cover crime scene examination (including photography and video) and audio examination/analysis, for example. However, the VFSC included all areas in the quality system uniformly, including three regionals and three other sites in the Melbourne central business district.

It was apparent as we worked through development and implementation that other Australian forensic agencies had a growing interest in accreditation and the implementation of quality management systems.

Representation through the Senior Managers of Australian and New Zealand Forensic Laboratories (SMANZFL) quickly led to the formation of a working party to develop an Australian Forensic Science Accreditation Program. Under the chairmanship of the Director of the National Institute of Forensic Science (NIFS), and from the extensive input of the National Association of Testing Authorities, Australia (NATA) and ASCLD-LAB, the Australian program is now a reality. Like ASCLD-LAB this program does not yet cover scene examination but the development of relevant criteria is well underway.

4 QUALITY MANAGEMENT SYSTEMS AND CONTINUOUS IMPROVEMENT PROGRAMS

The introduction to this paper mentioned a number of terms related to and involved with quality management systems (TQM) and CIP. High performance work systems is the American version of CIP and involves aspects of other terms such as benchmarking and best practice.

The process of benchmarking for best practice is a fairly logical, but not uncommon practice that is formalised in the titles describing it. What does benchmarking mean?

"An ongoing, systematic process to search for and introduce international best practice into your own organisation"[5].

"A method for organisational improvement that involves continuous, systematic evaluation of the products, services and process of organisations that are reorganised as representing best practice"[6].

The important elements are:

- that the process is ongoing and its application to quality management systems provides us with continuous improvement programs;
- it involves understanding the processes measured and how they can be implemented in one's own organisation; and
- it can involve comparisons with any organisations recognised as leaders in their field.

Formalising these processes such as quality systems implementation, accreditation and benchmarking leads to a systematic and ongoing program of self assessment, comparison and improvement. As the organisational environment continues to change ever more rapidly, comparisons with emerging best practices will allow us to continually improve and maintain a position of competitive excellence.

It serves to build and reinforce the change culture that in turn can improve service delivery within existing resource and policy constraints. The rigorous analysis of work practices encourages ownership and participation leading to employee empowerment and improved productivity. The process overcomes complacency and encourages change through identifying capability improvement opportunities.

Benchmarking however is not a stand-alone process and its linkage with other important management practices is essential.

5 CLOSING THE LOOP

Benchmarking can be conducted as an internal or external process. One element of an internal benchmarking exercise is audit. Earlier in this paper, we referred to key elements of quality systems as covered in ISO 9000[1] and ISO Guide 25[2] - the need for periodic audit to ensure continuing compliance of activities to the organisation's quality system is an important key element.

It is critical in the management of the CIP that the loop is closed and identified areas for improvement are addressed strategically and an assessment of their completion is made to formally sign them off.

Internal audits are the means by which the organisation monitors the proper implementation of, and adherence to, the quality system. At VFSC we have found the internal audit process most valuable and intend to extend it to involve an external component in 1995.

6 CONCLUSION

If we want to achieve and maintain good forensic science, we must have consistency and quality. The basis for this must be a comprehensive laboratory quality system that defines and documents operating parameters and encourages an organisational culture of continuous improvement.

7 REFERENCES

1. ISO 9000 "Quality management and quality assurance standards - Guidelines for selection and use" 1987.
2. ISO Guide 25 "General requirements for the competence of calibration and testing laboratories" 1990.
3. AS 2929 "Test methods - Guide to the format, style and content", Standards Australia 1990.
4. American Society of Crime Laboratory Directors - Laboratory Accreditation Board Manual 1994.
5. Benchmarking self help manual - your organisation's guide to achieving best practice 1993.
6. "Benchmarking and best practice: fashion or future?" - Royal Institute of Public Administration Australia Annual Conference 1994 (notes).

Accreditation and Quality Administration of Certified Reference Materials in China

Zhao Min and Han Yongzhi

NATIONAL RESEARCH CENTRE FOR CERTIFIED REFERENCE MATERIALS, BEIJING, PR CHINA

1. General Situation

Certified Reference Materials(CRM) are measuring standards with accurate value of quantity , widely used in quality assurance of measurement of products, drawing up and implementing social laws and regulations,etc. They are very important bases for internationalization of products, science and technology. Therefore many countries and international organizations pay full attention to the development, application and administration of CRM, set up special institutions one after another to direct, organize and coordinate the research and development, populization and application, carry out international coorperation in order to ensure the identity of CRMs among countries in the world.

Before 1949, the funding of P.R.China, CRMs were not used so many. The issue of spring steel used for worksite analysis in iron and steel factories by the iron and steel inspection committee of china in 1951 turned the history of research and application of CRMs to a new page. China started to develop CRMs not early, but with higher speed. By now there are 1350 kinds of CRMs passed the examination and accreditation of the State Bureau of Technical Supervision(SBTS), among them 816 kinds were approved as primary CRMs, 544 kinds as secondary CRMs including metallurgy, geology, building materials, environment, biology, food, nulear materials, petrochemicals, clinic analysis, chemical engineering, engineering properties, physico-chemical properties CRMs and etc. Table 1 will give a brief idea of the development of CRMs in china.

2. Accreditation of CRMs

2.1 basis and criteria of accreditation

CRMs in china are administrated unifiedly based on "the rule of CRMs administration" which was layed down according to the metrological law of china and refering to ISO guide 30,31,33 and 35. It is stipulated in the rule that the CRMs used in unifying value of quantiry consist of the CRMs for chemical composition analysis, the CRMs for measurement of physical properties and physico-chemical properties and the CRMs for measurement of technical properties; units which develop CRMs must have "licence for manufacture of CRMs; any new CRM must be identified for its grade and get relevant certificate.

The specification for affirming the grade of CRM, which was set in the rule for CRMs administration, are the fundamental basis for CRM quality accreditation. CRMs in china are classified into primary CRMs and secondary CRMs. The specification for primary CRMs are:

a) certified by an absolute measuring method or by accurate and reliable methods based on different principles or by multilaboratory measurement using one method if no other method could be fund for certification.

b) with the highest accuracy in china and the variation of homogeneity should not go beyond the scale of accuracy;

c) with good stability, its valid time should be longer than one year or reach to the international advanced level of same kind of CRM;

d) the packings should meet the request of "technical standard of primary CRMs";

The specification for secondary CRMs are:

a) certified by a comparative method to a primary CRM or by a method used in the certification of primary CRM;

b) accuracy homogeneity may not reach to the level of primary CRM but could content with demand of general measurement;

c) its valid time should be longer than helf year or meet the need of practical measurements;

d) the packings should accod with the demands of "technical standard for secondary CRMs".

2.2 Organization of accreditation

As metrological standards, CRMs in china are administrated by the State Bureau of Technical Supervision(SBTS). The CRM office set up in National Research Center for CRMs(NRCCRM) takes the responsibility on behalf of SBTS to coordinate administrate CRMs, carry out research plan and other quality management. CRM office is in charge of the accreditation. Meanwhile SBTS authorized CRM professional committee(RMPCO) which is one of the sub-organizations of "china metrology and measurement society" to perform the examination and evaluation of CRM technical grade. The accreditation system of CRM is shown in Fig.1.

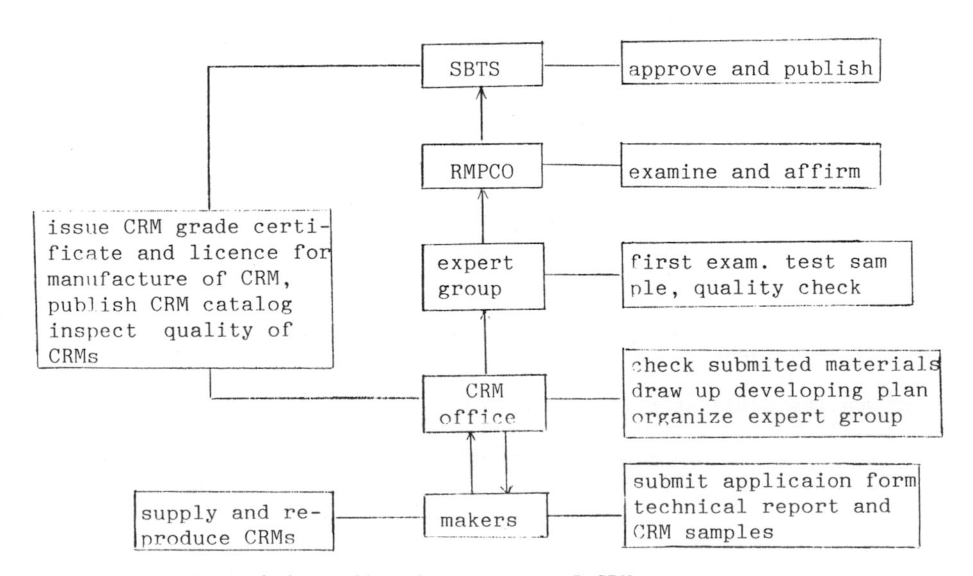

Fig. 1 Accreditation system of CRMs

2.3 Procedure of technical evaluation

a) Application

A primary CRM must have a "grade certificate of primary CRM" and a "licence for manufacture of primary CRM". So makers who intend to apply for these should ask CRM office in NRCCRM, fill in application forms printed by SBTS, submit the form with the comments and signature of up level institute. together with some CRM samples and the following documents:

° assignment of developing CRM;

° technical report on CRM development including preparing procedure testing method for homogenity, techniques and results of certification, dataprocessing, stability investigating and etc.;

° comparison of properties with same kind CRM abroad or at home;

° trial report;

° certificate of CRM (drift);

° abilities and ways for CRM supply;

° status of manufacture including facilities, analytical instruments, laboratory conditions, technical persons, etc..

b) grade evaluation

The CRM office, after checking application documents and samples, invites 3 to 4 experts who are familiar with the CRM to be examined from universities or high level research institutes to appreciate the materials submitted based on the specification of primary CRM, and write down their opinion and suggestion after discussion. If there are items needed to recheck by experiment, the CRM office would ask some authoritative laboratories to do, then synthesizing the experts suggestion and recheck results give the applicant a notice of the first evaluation in written form.

The CRM passed the first evaluation will subject to the second or final evaluation of "CRM professional committee". It usually takes reply ways. The committee members(39 members come from universities and research institutes) ask question on the CRM developed, the applicant gives answer. Finally the committee makes decision by vote.

c) Issue certificate and licence

The CRM office reports the voted result to SBTS for approval. The approved CRM will be numbered,put into CRM catalog and published; meanwhile SBTS would issue a "grade certificate of primary CRM" and a "licence for manufacture of primary CRM".

The procedure for accreditation of secondary CRM is same as that for primaryCRM; the only difference is the manage department of ministries which the CRM manufactures belong to take the place of CRM office to carry out the evaluation. Passing the final evaluation the secondary CRM will be reported to SBTS for approval and get "grade certificate of secondary CRM" and licence .

In china primary CRM is approved only one of each kind, secondary CRM may have more for their wide use and large amount consumption.

3. Quality administration of CRMs

As metrological standards CRMs transmit values of quantity. The transitive system for value of quantity of CRMs in china is shown in Fig.2.

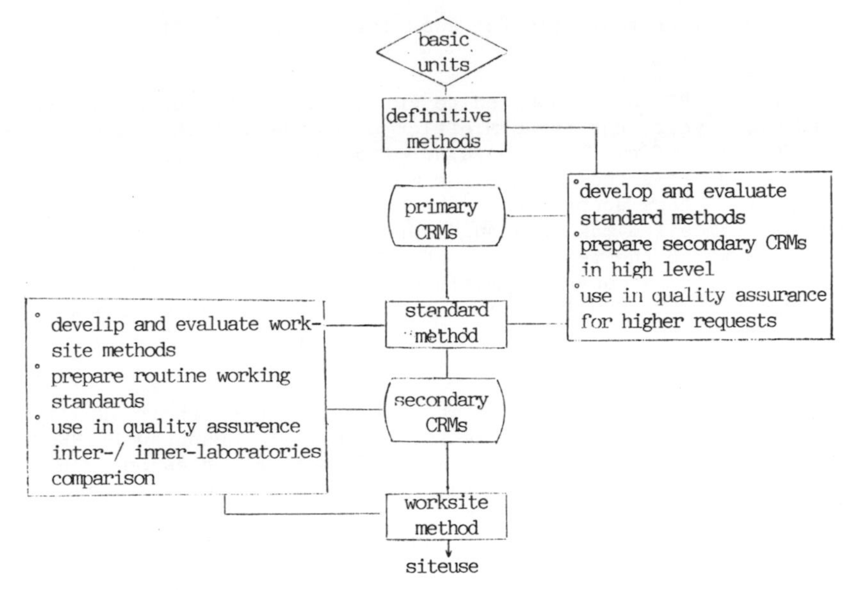

Fig.2 transitive system for value of quantity of CRMs

Since CRMs are widely used as standards of material measures to check or calibrate apparatus, assess measuring methods, assign values of materials and products in industrial measurement, quality control, environmental monitoring, clinic analysis, scientific research etc., the quality of CRMs are extremely important, so great efforts were made in china for CRMs quality administration.

3.1 Lay down "technical standard for primary CRMs"

In order to develop CRMs and ensure the quality of CRM produced by different makers SBTS layed down "technical standard for primary CRMs". In the standard principles and procedures which may followed in developing and preparing CRMs were described, for instance, how to select condidate, sampling method, minimum sample number for homogeneity test, measuring methods, results evaluation, stability investigation and etc. To ensure the accuracy and reliability of value of quantity it stipulates in the standard that one definitive method or two methods with different principles should be used in the certification, multilaboratory measurement could be used also for no other suitable method. The standard emphasizes that all metrological instruments must be verified by metrologocal institutions, the traceabilities of value of quantity must be assured, etc. Makers as long as carry out their research following this technical standard, the CRMs they developed would be in high quality.

3.2 Strict evaluation

The CRMs quality accreditation is carried out in two steps of the first evaluation and second evaluation based on "the rule of CRM administration". Any problem fund in the first evaluation must be solved before going to the second evaluation. The experts and committee members with top level of theoretical foundation and plentiful experiences could point out problems and means to solve them, so that two times evaluation would play very improtant role for ensuring the quality of CRMs.

3.3 Check at any time

The makers of CRMs are requested to submit stability report of CRM developed every year, and the CRM office often check technical properties of CRMs to comfirm that the value of quantity are still correct and reliable. Besides the licence for manufacture of CRMs should be renewed every 5 years according to the rule of CRM administration. When the day comes the makers should apply for a new licence. The CRM office will examine the test report of the CRM in 5 years and other materials submited by makers, sometimes few samples will be analyzed to make sure that the CRM is in good condition. After these a new licence can be issued.

3.4 Changing informations

In order to promote the development of CRMs, rise up their quality a technical information network was set up in 1991 sponsored by NRCCRM. Now there are 205 members in it. Through conducting correspondence and training courses of CRM, distributing the prints of "developments in research of CRMs", opening a special column in J. CHINESE TECHNICAL SUPERVISION, and etc. makers could change information, learn from each other. the quality of CRMs is improved fast. In addition, NRCCRM set up COMAR data base to provide information of CRMs developed in other countries. All these make the development of CRM more active and in higher level.

Table 1. The development of CRMs in China

year / grade / amount / category	1985	1986	1987	1988	1989		1990		1991		1992		1993		1994	
	(1)	(1)	(1)	(1)	(1)	(2)	(1)	(2)	(1)	(2)	(1)	(2)	(1)	(2)	(1)	(2)
A	37	39	56	73	98	0	116	8	131	8	137	8	147	9	172	9
B	10	22	28	48	50	0	66	0	66	0	95	0	96	0	126	0
C	0	0	0	0	11	0	17	2	17	2	23	2	26	2	35	2
D	16	16	16	16	47	7	57	7	59	7	83	7	92	7	97	7
E	0	0	0	0	0	0	0	0	0	3	0	3	0	3	2	3
F	0	0	3	4	4	0	5	17	6	30	16	23	16	66	22	88
G	12	16	22	59	72	0	74	0	123	17	128	23	143	23	159	36
H	12	16	30	40	44	59	47	114	52	157	65	187	81	203	90	238
I	0	0	0	0	1	1	8	2	3	23	8	5	15	6	30	12
J	0	0	0	0	0	0	0	0	0	0	0	0	0	0	5	0
K	0	0	5	6	6	0	6	0	14	0	14	0	14	0	14	0
L	0	0	0	0	0	8	0	8	0	8	0	8	0	9	3	12
M	11	11	12	12	40	13	50	65	52	77	53	77	53	104	61	128
total	98	120	172	261	373	88	446	223	529	312	622	344	683	432	816	544
increase/year	98	22	52	85	112	88	135	83	83	89	97	22	61	88	133	112
					200		208		172		119		149		245	

note:

A: ferrous metals
B: non-ferrous metals and gases in metals
C: building materials
D: nuclear and radioactivity
E: polymeric materials
F: chemical products
G: geology
H: environmental
I: clinic chemistry, biomedical and pharmaceuticals
J: food
K: energy resources
L: technological and engineering
M: physics and physico- chemistry

Customer Satisfaction: A Reality in Analytical Laboratories?

Anne Nicholson

ANNE NICHOLSON ENTERPRISES PTY LTD, 32 MILLICENT STREET, ROSANA, VICTORIA 3084, AUSTRALIA

1.0 INTRODUCTION

In today's economic climate the organisation that demonstrates it can meet and exceed its customer's needs and expectations accurately and on time is the one that will succeed. This organisation is characterised by its commitment to customer focused strategies which facilitate a high level of trust and confidence between itself and its customers as demonstrated through a close working relationship with its customers. The argument being offered is that this loyal customer base is not an accident. It results from hard work and an organisational structure which promotes the rights of all of its customers, internal and external. Judgements will be suspended and customers will be actively listened to. To know your customer well, means there is no competition. Is this your vision for the future? The aim of this paper is to discuss some of the key issues relating to understanding your customers needs.

As a laboratory do you know who your customers really are? Are they, as in the example of the Pathology Laboratory, the hospital or medical practitioner who orders the test or the patient who pays the actual account? What about those laboratories where the actual laboratory staff never come into contact with the person who pays for the service. Somehow the customers' needs have to be translated into laboratory jargon or functional terms and transmitted to those who will perform the actual work. A structure which creates barriers between the customer and the actual delivery of the required service. A process which frequently leads to frustration and ultimately complaints from both the staff and external customers.

1.1 Commitment to Customers in Analytical Laboratories

The level of commitment to customers can only be determined through a deliberate review of your customers perceptions and attitudes towards your laboratory. This would entail reviewing staff attitudes towards customers as well as deliberately asking customers for their perceptions of the degree to which the organisation values their feedback and input.

Check yourself, is there a customer culture within your Laboratory? As a beginning just ask yourself and your staff these questions :

- Is there a belief that customers have unrealistic expectations?
- Is there a belief that customers cannot understand the work of the Laboratory?
- Is customer feedback actively sought and findings and changes in practices fed back to customers?

- Do all staff know who their customers are and what their specific needs and expectations are?
- Are there clear lines of communication between departments?
- Does the organisational structure provide an interface between departments or does it create barriers?
- Is there a feeling of powerlessness with regard to controlling the work?
- Is the impact of management decisions on individuals always considered prior to committing to the decision?
- Are staff recognised for work that has been done well?

An affirmative response to these questions by yourself and all staff would indicate the existence of a customer culture. On the other hand negative responses suggest the need for a refinement of your approach to your customers. The above questions are important as they highlight the importance of actively working with staff and customers to achieve agreement as to what constitutes a quality service.

2.0 CUSTOMER SATISFACTION : A DILEMMA

The argument being presented in this paper is that laboratories could work smarter if they took the time to develop insight into the differences in perception between themselves and customers as to the quality and priority of product and service levels. This insight will enable them to focus their energy and resources on satisfying their customers actual priority needs and expectations rather than on their perceptions of what they believe would satisfy their customers.

The potential for a difference in perception between customers and suppliers was highlighted by an important study into airline passengers' level of satisfaction with the service. This study concluded that there were marked differences in concerns between management and passengers (1).

Their findings were that passengers' priority concerns included the following :

- the airline should be responsive to lost bags
- information of flight delays should be made available immediately
- an expectation that bags would be handled carefully

On the other hand it was found that the airline managers' priority concerns included :

- earning a profit
- ensuring courteous cabin crew
- flights departing on time.

On the basis of study findings it is clear that working on correcting the airline managers specific concerns would not lead to increased passenger satisfaction. Despite its efforts the airline could have missed achieving its goal - increased market share - because it has overlooked a fundamental business reality that increased market share is closely correlated to customer satisfaction.

The importance of this study for laboratory personnel lies in its message to check your assumptions about your customers and their expectations prior to committing time and resources to introducing a new initiative or modifying existing services and/or products.

2.1 Determining Customer Satisfaction : an approach

The outcomes from the previous study were fully supported by a Continuous Quality Improvement Pilot Study that was undertaken at the Austin Hospital, Melbourne in 1993/4 (2).

This study was undertaken with a view to developing a customer focused culture in a clinical health care setting.

Prior to the commencement of the study a customer satisfaction survey had been undertaken. Staff expressed disappointment with the results. They felt that although the information generated through the survey indicated that respondents were generally happy with the service they failed to gain any significant information with regard to how they could improve their services to their customers.

The actual questionnaire was reviewed and it was found that the questions focused on respondents' level of satisfaction with the existing service. There was no customer input into the survey or questionnaire design, nor was any attempt made to accurately identify and define customers' priority concerns as a baseline to designing the survey.

Given the identification of the limitations in the design of the initial satisfaction survey the whole survey process was redesigned to ensure a strong customer focus. The goal was to ensure the questions put to customers reflected their real and perceived needs and expectations.

2.2 Methodology

The survey was totally redesigned to enable clinical staff to isolate and define specific customer satisfaction criteria. The service delivery process was reviewed by means of process flow charting. The charts were analysed to define the effectiveness of the existing service delivery systems and to isolate service delivery barriers. This analysis enabled staff to develop insight into the potential impact of work practices on their customers. Customer complaints were then reviewed and analysed against the outcomes from the work systems reviews. This process provided staff with a high level of insight into the strengths and weaknesses of their actual work practices. They then prepared a series of broad questions that they could put to their key customers and focus groups. These questions enabled them to test their assumptions about customers perceptions as to what constituted an acceptable level of clinical service. The questions were designed to specifically isolate their customers priority needs and expectations.

The data generated through the interviews was analysed and translated into a questionnaire for distribution to a targeted population. This population included those who had responded to the original satisfaction survey.

2.3 Findings

The importance of this process can be seen in the results. The feedback from the respondents identified significant service deficiencies. Deficiencies which the initial survey totally missed. The information forced them to re-assess their work practices and redefine their service delivery priorities. The information also provided them with a baseline for improving their market competitiveness. Specifically the findings included :

- the department was described as not performing very well; however it was said to be performing as well as its competitors
- specific opportunities for improving the service were identified; these included :
  - a need for written information describing the range and prices of products
  - the availability of written information describing the range of services offered
  - clarification of how to access the services and products

The value of the methodology for ensuring an appropriate and effective survey design can be appreciated through reference to the following tables.

Table 1 *Strategic Map, Importance vs Performance*
The Strategic map is a profile demonstrating the relationship between the variables in regard to importance and performance.

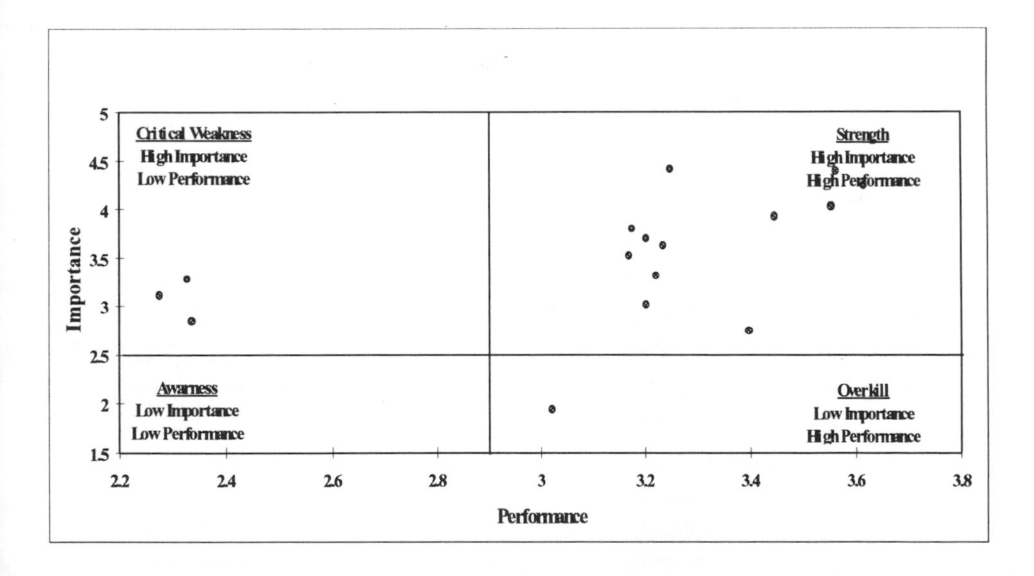

Critical Weakness : *generally these items need to be improved in the short term for the growth of the department in order to continue moving ahead.*

Strength : *performing well.*

Awareness : *this may be due to lack of awareness.*

Overkill : *It may be necessary for items falling into this segment to confirm their real importance, is there to much effort put into them?*

Table 2 *Importance vs Performance Chart*
This chart compares the customers perception the current performance of service and the level of importance they place on it.

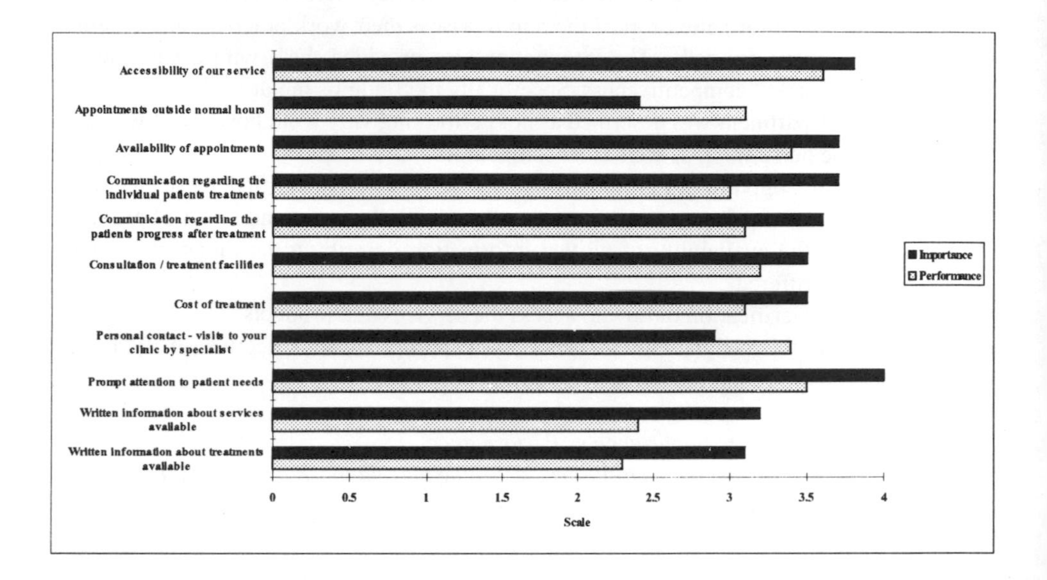

Table 3 *Improvement Needs Ranked - the gaps defined*
Negative results indicate areas where the performance is not meeting importance, while positive results indicate areas which are more than meeting customer needs.

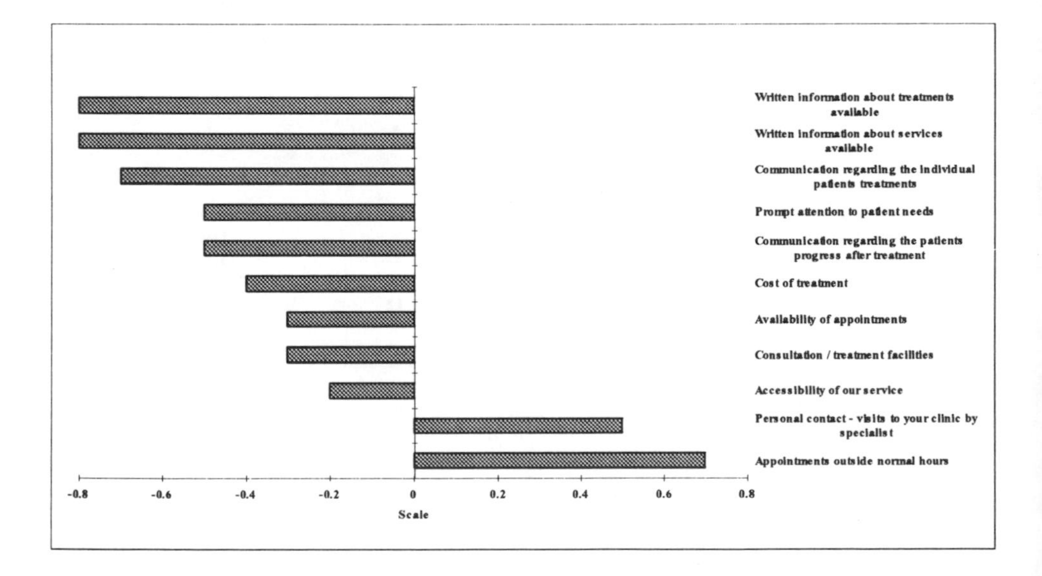

Table 4 *Performance vs Competition Chart*
Customers perception of where you sit in regard to your competitors.

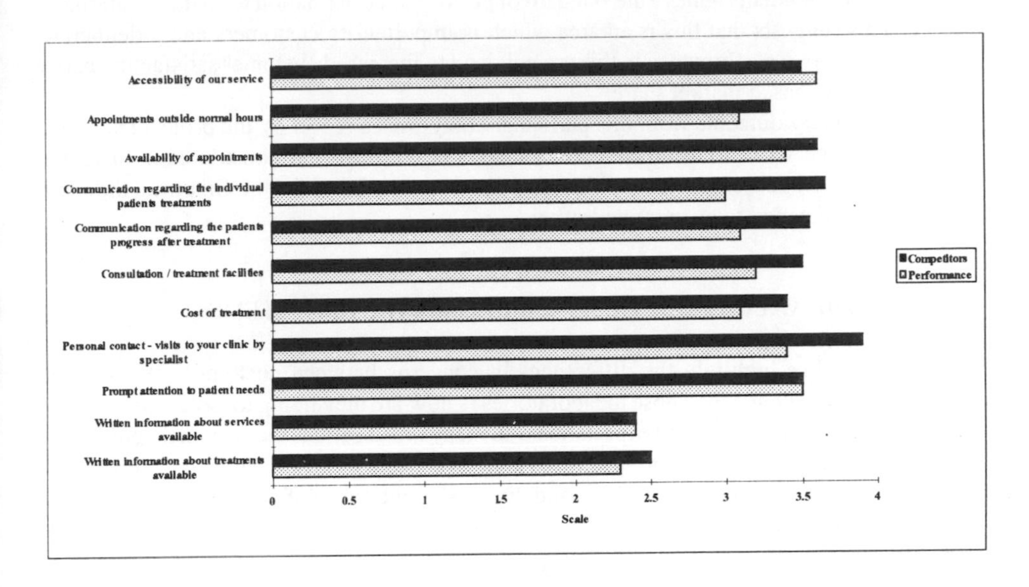

Table 5 *Improvement Needs Ranked - Key Indicators Identified*
The more negative a result the bigger the gap between you and your competitor, while positive results indicate area where you are ahead of your competitors as judged by your customers.

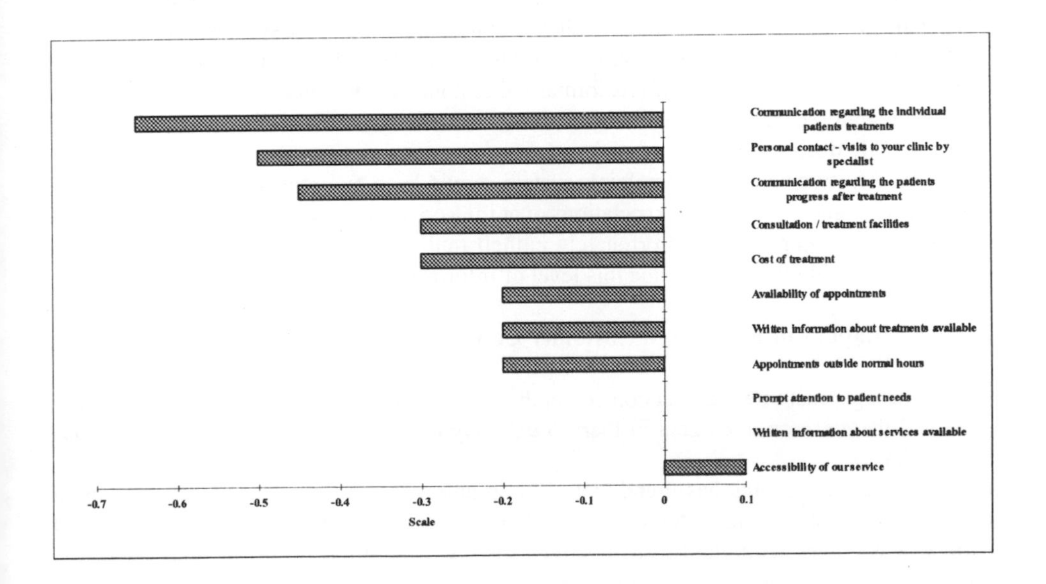

Table 3 indicates that the area which customers find the most lacking is 'written material regarding services and treatments'. However Table 5 shows that when compared with competitors respondents believe the standard of performance in relation to written material is similar. This suggests that this is an area which is important to customers and a deliberate strategy to improve performance in this are will lead to increased customer satisfaction and a potential for increased market share.

A secondary outcome from this particular study was a report by the project team that they had learned what a customer focus actually meant and what constituted a customer focused service. Their customers on the other hand reported that for the first time they felt they had been listened to by an organisation.

3.0 THE VALIDITY OF CUSTOMER COMPLAINTS SYSTEMS

The above studies highlight the differences in concerns between customers and service suppliers. The challenge they pose to laboratories is how are they going to isolate and clarify their specific customer needs and expectations? In many organisations the customer complaints system acts to trigger responses to specific customers concerns. Unfortunately it is a relatively crude method for accurately identifying and diagnosing the cause of customer concerns and the overall number of complaints. This limitation can be minimised if it is fully understood and used accordingly.

3.1 Requirement For Formalised Complaints Systems

For example, most analytical laboratories will have formal customer complaint systems. These systems are a requirement for laboratories seeking accreditation through NATA and certification to the ISO 9000 : 1987 or the AS/NZS ISO 9000 : 1994 quality standard series. The standards and ISO/IEC Guide 25 1982 tend to emphasise work systems and practices, that is an internal rather than an external customer focus (3). The two audit systems require laboratories to implement responsive, formalised customer complaints systems. These systems must be monitored, reviewed and modified to ensure customers' complaints are adequately and promptly dealt with.

An effective customer complaints system should keep an organisation fully informed about its customers needs and expectations or at the very least provide an accurate picture of prevailing customer trends. Unfortunately in their traditional form they are not sufficiently sensitive to enable them to provide this level of information.

3.2 Limitations of Formalised Customer Complaints Systems

Through customer satisfaction research it has been found that for every complaint received there are on average a further 26 unhappy customers, six of whom have serious complaints.

This same research has found that if an organisation commits itself to maintaining a close relationship with its customers it can retain its customers. This is based on the fact that where customers' complaints are resolved 70% will continue to do business with the organisation. If the problem is resolved very quickly, 95% of those who have complained will continue to use that organisation (4).

The above findings must be considered against the actual probability of customers complaining about perceived poor service. The research findings suggest that on average 96% of unhappy customers never complain to the organisation. The implications for an organisation are far reaching. Even a minimal commitment to diagnosing and identifying customers' key concerns and matching these to the organisation's direction and working actively with customers to achieve open two way communications will have a positive, direct impact on viability and market share (5).

Laboratories that have committed themselves to the certification and accreditation processes have demonstrated their commitment to providing a quality customer service as well as extending their market share. On the basis of customer satisfaction research then it would appear to be in their interest to develop and initiate a customer service strategy that extends the requirements of existing laboratory service standards. This recommendation is based on the premise that the formal complaints currently being received are likely to be simply the tip of the iceberg.

4.0 DEVELOPING A CUSTOMER SERVICE STRATEGY

Given the complexity of the services provided by analytical laboratories and the reliance on expert and highly specialised skills, knowledge and counselling the achievement of a customer focused culture can only be realised if the impact of the system on a customer can be understood from their perspective.

A useful model for developing insight into the relationship between work systems and practices and customer satisfaction is the Quality Benchmark Deployment Model. This model focuses on understanding the impact of actual work systems and practices on customers and provides a means for agreeing to performance measures that are based on customer satisfaction (6).

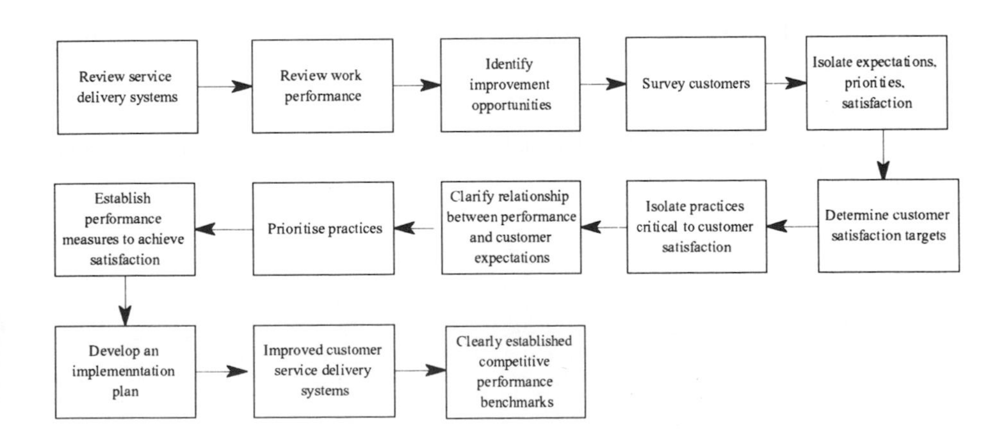

Figure 1 *Clarification of the Relationship Between Customer Satisfaction and Work Systems and Practices*

It is important to understand customer complaints in context. This model offers both a template for achieving this level of understanding as well as providing a systematic process for clarifying and analysing customer needs and expectations.

7.0 CONCLUSION

Customer feedback and complaints can be an important tool for improving a laboratory's competitive position. The reality is that most organisations do not know how to effectively manage complaints and run the risk of losing customers.

Complaints arise out of differing priorities and levels of concerns between organisations and their customers. The challenge for laboratories is to view customer complaints as the tip of the iceberg and to develop positive customer satisfaction processes. Processes that will result in a customer focused laboratory culture which encourages its customers to act as partners in the process. Through concentrating on your customers and treating them as legitimate partners in your laboratory you will find there is no competition and you will have a loyal customer base.

References

1. Kloppenborg T, Gourdin T, 1992 Up in the Air About Quality QUALITY PROGRESS Vol xxv, No 2, p31

2,7. Nicholson A, 1994 Listening to Our Customers - A Quality Benchmark Deployment Strategy in QUALCON 94 BOOK OF PROCEEDINGS Melbourne, Australian Organisation for Quality

3. Locke J W, 1993 Quality Standards for Laboratories QUALITY PROGRESS Vol 26, No 7, p91

4,5. Albrecht K, Bradford L, 1990 THE SERVICE ADVANTAGE USA, Dow Jones-Irwin

6. Swanson R, 1993 Quality Benchmark Deployment QUALITY PROGRESS Vol 26, No 12, p84

The Journey to Quality and Continuous Improvement: The First Few Steps

Carl A. Gibson and Len R. Stephens

VICTORIAN INSTITUTE OF ANIMAL SCIENCE, 475 MICKLEHAM ROAD, ATTWOOD, VICTORIA 3049, AUSTRALIA

1 INTRODUCTION

The Victorian Institute of Animal Science (VIAS) is one of the major Institutes of the Victorian Government's Agriculture Victoria. VIAS is responsible for the provision of a wide range of research and diagnostic services to a broad customer base in the livestock industries.

VIAS has evolved from a number of independent units of the Department over the last five years. In 1990 the Veterinary Research Institute (VRI) located on two campuses combined to a single location on the northern outskirts of Melbourne (Attwood). This was followed in 1991 by the amalgamation of the VRI with the Animal Research Institute (Werribee) to form VIAS on two campuses at Attwood and Werribee, and in 1992 by the relocation of the Livestock Improvement Unit to Attwood. Through this process, staff numbers have decreased through natural attrition and voluntary departures from over 500 to approximately 200.

In 1994, VIAS underwent a structural reorganisation to form four Divisions: Animal Health (AH), Animal Production and Welfare (APW), Molecular Biology and Genetics (MBG) and Institute Services (IS) (Figure 1).

In conjunction with these significant internal shifts, the Institute's environment has also been subjected to significant turbulence. The introduction of wide-reaching workplace reforms within the public service and turbulence within the agricultural industries, have all resulted in significant effects on the functions of the Institute. Global changes such as the GATT agreements, the European Community's trade and quality policies and the quality movement within the US Department of Agriculture and the Food and Drug Administration are placing mounting pressures on the quality requirements for Australian agriculture. In recent years large segments of Australian agriculture have begun to change and embrace many of the quality philosophies and adopt the principles of best practice.

This has begun to have serious implications for the core activities of VIAS which include laboratory and field research into novel vaccines, genetic improvement of livestock, livestock behavioural studies, disease diagnosis and the health certification of livestock exports.

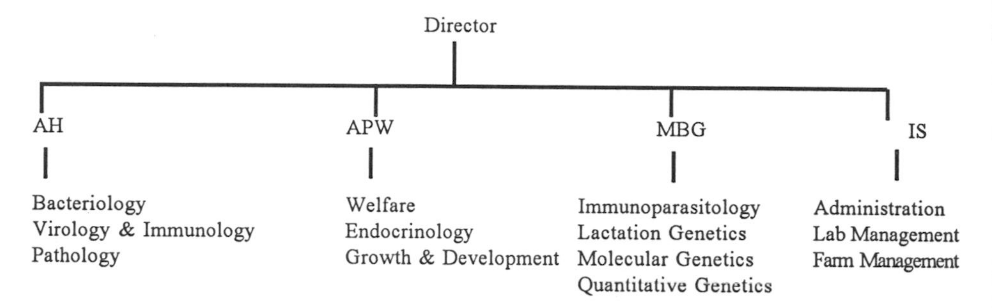

Figure 1: *Divisional and Departmental Structure of the Victorian Institute of Animal Science*

In late 1993 it was acknowledged that the way in which VIAS conducted its business would have to improve if the Institute were to remain competitive in attracting funding for its activities. A decision was made that the future survival of the VIAS would be dependent upon the successful introduction of a quality culture into the organisation.

2 THE EARLY STAGES OF INTRODUCING TOTAL QUALITY MANAGEMENT

Through the 1990's a number of operational Departments of VIAS began to informally introduce some of the quality philosophies into laboratories. This was initially restricted to the diagnostic and analytical laboratories and included the use of statistical process control techniques, participation in interlaboratory proficiency programs, limited staff training in quality techniques and the monitoring of some of the key outputs of these laboratories.

The boost for the introduction of TQM into the whole of VIAS came at the start of 1994 with the expressed commitment of the Institute's Director to the concepts of continuous improvement and a strong desire to see management and staff embrace its principles. This initial step involved a series of meetings involving staff at all levels, at which the vision for the future direction of VIAS was expressed.

Following this, a Department Head with a background in the quality movement was given the responsibility for developing quality management systems for VIAS. The next major step involved an assessment of the current status of VIAS in a number of quality related areas.

3 THE ASSESSMENT OF ORGANISATIONAL CLIMATE

3.1 Performance Measures

There have been a myriad of text books published on TQM, resulting in a wide range of definitions of quality and of areas to focus on in order to develop a quality organisation. A simple model for quality organisation was adopted early in the VIAS TQM program, based on three corner stones: leadership, at all levels of the organisation; people, all of our staff; and customers, both internal and external (Figure 2).

The initial assessment involved an examination of a range of operational parameters of the Institute including: ratios of senior to junior scientists and technicians, relevant staffing ratios of administration, technical and scientific staff, the ratio of operational staff to administrative in other units of Agriculture Victoria and the annual rates of staff turnover. Other measures of the staff environment involved occupational health and safety trends, monitoring annual improvements in injuries, Workcare claims and incident days lost, and detailed investigations into high risk injury areas. Trends in financial measurements such as income ratios, output ratios and the unit costs of producing scientific publications were mapped back for a minimum of three years.

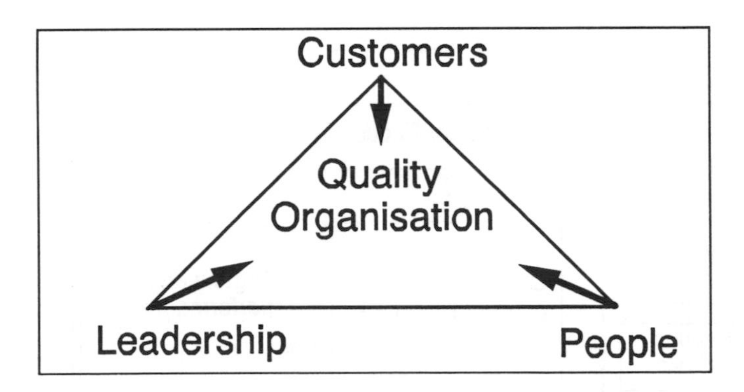

Figure 2: *The Cornerstones of a Quality Organisation*

During the assessment process, all Department Heads participated in a strategic planning workshop during which laboratory functions and core competencies were defined and a wide range of customers were identified and segmented (Table 1).

3.2 Staff Attitude Surveys

This information, however, only provided a partial picture of the status of VIAS. The real backbone of the assessment was provided by the use of a confidential attitude survey, in which staff at all levels of the organisation were encouraged to participate. The survey used was a graded answer questionnaire and explored staff responses to a range of issues which were relevant to the criteria of the Australian Quality Awards (Figure 3). The survey clearly identified a range of problem issues for staff, which indicated the need for improved leadership, whilst also indicating that staff took pride in their work and recognised the importance of providing a customer-based service.

Issues raised by the use of the survey were subsequently followed up with a series of informal interviews and discussion groups with a representative sample of staff members. It became apparent through this process that, although staff believed they fully understood their customer needs, few had actually asked their customers what they in fact needed.

3.3 Customer Satisfaction

A number of approaches were used to better define customer needs, ranging from informal one-to-one meetings with customers to bringing customers into the laboratories for educational programs (eg farmer or veterinarian groups).

However, it was the determination of customer satisfaction that had the greatest impact on staff. A questionnaire was designed, based upon a range of needs identified by clients over a number of personal meetings conducted over the previous months. Users of the diagnostic services each received the questionnaire in the post with a prepaid return envelope. The questionnaire required the customer to rate i) the performance of VIAS in 20 specific areas against a liner scale from 1 (poor) to 5 (excellent); ii) the performance of VIAS relative to other laboratories; and iii) the relative importance of each of the needs to the customer.

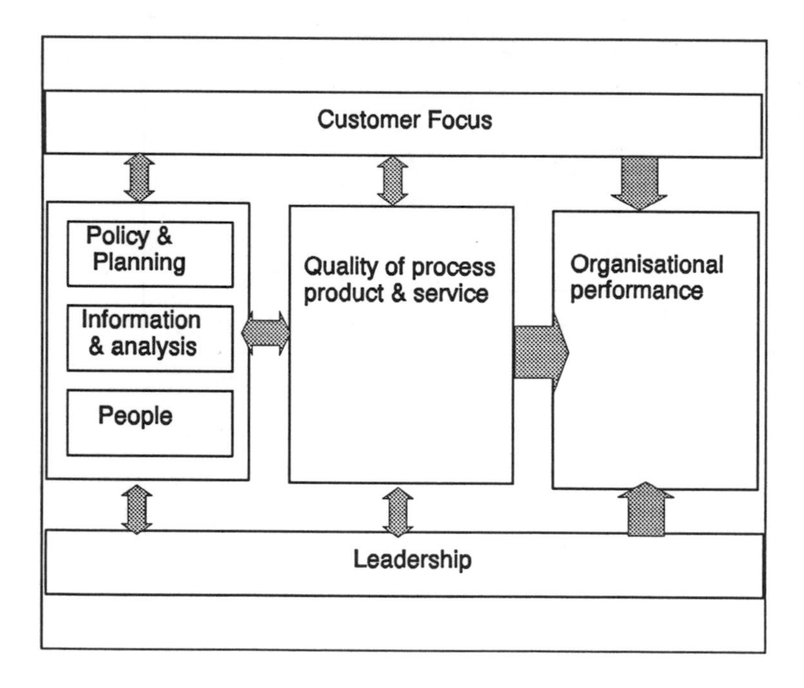

Figure 3: *The Australian Quality Awards Criteria used for Assessment of VIAS Culture and Climate*

Table 1: *Customer Groups Serviced by each of the VIAS Divisions*

Customer group	**AH**	**AP**	**MBG**	**IS**
Agriculture Victoria internal customer	XX	X	X	XXX
RIRFS	XXX	XXX	XXX	
Vic. Government Agencies	XXX	XX	XX	
Federal Government Agencies	XXX	XX	XX	X
Pharm/chemical industries	XXX	XX	XXX	
Rural/farm businesses	XXX	X		
Veterinary practitioners	XXX			
Pathology labs	XXX			
Universities	XX	XX	XXX	
Interstate Government agencies	XX	XX	XX	
Hospitals	X			
Overseas trading partners	XXX			
Breed Associations	XX		XXX	

+ the number of X's indicates the relative quantity of service provided

The customer survey achieved a good response rate of over 70%, with the results providing a number of surprises. Laboratory functions that were believed to be of good quality were often seen as disappointing to their customers and in a number of cases were significantly poorer than those services provided by competing laboratories (Table 2). On the other hand, areas where laboratory staff believed increased resources and extra effort were required proved to be of relatively low importance to the customers, for example extended service hours or the provision of new specialised test packages.

3.4 The Final Assessment Process

The range of data and information collated was subsequently used to form the basis of an evaluation against the criteria of the Australian Quality Awards (AQA) undertaken by an Australian Quality Council trained internal assessor. This process revealed both strengths and opportunities for improvements for the organisation. The process also resulted in a qualitative score of the organisation against each of the AQA criteria. VIAS achieved a total score of 179 out of 1000 points at the start of its quality journey. However, through the gradual introduction of quality procedures, the organisation has improved to over 400 points in under 12 months of TQM activity.

Table 2: *Customer Survey Results Identifying the five Poorest Aspects of VIAS Diagnostic Services Compared to Other Laboratories*

Importance to customer	Customer Needs	Rank	Competitive gap+
4th	Prompt notification of results	20th	-13.26
3rd	Overall ease of use of service	19th	-12.10
1st	Overall quality of services	18th	-11.95
5th	Provision of specialist advice	17th	-11.64
2nd	Reliability	16th	-11.63

+ The competitive gap was determined as:

(the customer importance score) X -(5 - the competitive performance score)

4 OUTCOMES OF THE ASSESSMENT PROCESS

The assessment process provided a comprehensive snapshot of VIAS and identified the Institute's strengths and areas of opportunity for quality improvement activities. Areas that were identified for improvement activities included:

- The establishment and training of a TQM implementation team: Training was undertaken using outsourced providers and internal expertise. Knowledge was subsequently `cascaded' through the organisation.

- Team building courses for managers and staff: Both external and internal trainers were used, the training centred around `experiential' and `comfort zone' formats.

- Process improvement: All groups have defined and commenced the monitoring of key performance indicators in critical results areas (what gets measured gets done!). The next stage is to incorporate more relevant customer-defined indicators.

- Customer service improvement: All staff were invited to join a customer focus group, with each group serving a specific industry customer group. This customer focused structure was then incorporated into a matrix overlaying the traditional discipline based structure.

- Communication improvement: Individual groups have begun to identify barriers to both upward and downward communications. This issue has proved to be a critical one for the whole of Agriculture Victoria, which is undertaking a Department-wide study into communication needs.

- Occupational health and safety improvements: OH&S became part of the assessment criteria for assessing staff performance for pay increases. All Department Heads were given specific corporate safety responsibilities and accountabilities.

- Empowerment of staff: This has started with educating managers in the role of coach and facilitator. Staff at all levels participate in customer focus teams and process improvement teams.

5 CONCLUSION

The quality journey that VIAS has entered upon is still in its early stages and as an organisation we are learning daily. In line with many of the case studies in the literature we have made mistakes, and with this experience we now recognise the necessity of a few key issues which are vital to the success of a TQM program.

- It is essential that the chief executive officer is behind the process, in heart in mind and in action.
- All managers must "walk the talk", ie be visible in deeds as well as in words.
- Fear must be removed from the organisation.
- Staff at all levels must see that their opinions are listened too.
- Everyone serves a customer; if they do not, then they help someone who does.
- The customer is of prime importance, but the other stakeholders cannot be ignored (eg suppliers, community, staff, industry, owners, boards of directors or government).
- The end is never reached. It truly is a journey of continuous improvement.
- Staff must be committed and realise the value of what they are doing.
- In the beginning it will require a great effort to commit resources. Freeing up time is everyone's greatest problem.
- The concept that `quality is not an add-on, but the way we do business' must be driven from day one.

SELECTIVE LIST OF RELEVANT

INTERNATIONAL STANDARDS AND GUIDES

03.120 Quality

03.120.10 Quality management and quality assurance

ISO 8402:1994 *Ed. 2* *39 p. (R)* *TC 176 / SC 1*
Quality management and quality assurance — Vocabulary
Trilingual edition

ISO 9000-1:1994 *Ed. 1* *18 p. (J)* *TC 176 / SC 2*
Quality management and quality assurance standards —
Part 1: Guidelines for selection and use

ISO 9000-2:1993 *Ed. 1* *16 p. (H)* *TC 176 / SC 2*
Quality management and quality assurance standards —
Part 2: Generic guidelines for the application of ISO 9001, ISO 9002 and ISO 9003

ISO 9000-3:1991 *Ed. 1* *15 p. (H)* *TC 176 / SC 2*
Quality management and quality assurance standards —
Part 3: Guidelines for the application of ISO 9001 to the development, supply and maintenance of software

ISO 9000-4:1993 *Ed. 1* *21 p. (L)* *TC 176 / SC 2*
Quality management and quality assurance standards —
Part 4: Guide to dependability programme management
Bilingual edition

ISO 9001:1994 *Ed. 2* *11 p. (F)* *TC 176 / SC 2*
Quality systems — Model for quality assurance in design, development, production, installation and servicing

ISO 9002:1994 *Ed. 2* *10 p. (E)* *TC 176 / SC 2*
Quality systems — Model for quality assurance in production, installation and servicing

ISO 9003:1994 *Ed. 2* *7 p. (D)* *TC 176 / SC 2*
Quality systems — Model for quality assurance in final inspection and test

Technical Corrigendum 1:1994 to ISO 9003:1994
Ed. 1 *0 p.* *TC 176 / SC 2*

ISO 9004-1:1994 *Ed. 1* *23 p. (L)* *TC 176 / SC 2*
Quality management and quality system elements —
Part 1: Guidelines

ISO 9004-2:1991 *Ed. 1* *18 p. (J)* *TC 176 / SC 2*
Quality management and quality system elements —
Part 2: Guidelines for services

Technical Corrigendum 1:1994 to ISO 9004-2:1991
Ed. 1 *0 p.* *TC 176 / SC 2*

ISO 9004-3:1993 *Ed. 2* *21 p. (L)* *TC 176 / SC 2*
Quality management and quality system elements —
Part 3: Guidelines for processed materials

ISO 9004-4:1993 *Ed. 2* *25 p. (M)* *TC 176 / SC 2*
Quality management and quality system elements —
Part 4: Guidelines for quality improvement

Technical Corrigendum 1:1994 to ISO 9004-4:1993
Ed. 1 *2 p.* *TC 176 / SC 2*

ISO/DIS 10005 *Ed. 1* *TC 176 / SC 2*
Quality management — Guidelines for quality plans (formerly ISO/DIS 9004-5)

ISO/DIS 10007 *Ed. 1* *14 p.* *TC 176 / SC 2*
Quality management — Guidelines for configuration management (Formerly DIS 9004-7)

ISO 10011-1:1990 *Ed. 1* *7 p. (D)* *TC 176 / SC 2*
Guidelines for auditing quality systems —
Part 1: Auditing

ISO 10011-2:1991 *Ed. 1* *5 p. (C)* *TC 176 / SC 2*
Guidelines for auditing quality systems —
Part 2: Qualification criteria for quality systems auditors

ISO 10011-3:1991 *Ed. 1* *3 p. (B)* *TC 176 / SC 2*
Guidelines for auditing quality systems —
Part 3: Management of audit programmes

ISO 10012-1:1992 *Ed. 1* *14 p. (G)* *TC 176 / SC 3*
Quality assurance requirements for measuring equipment —
Part 1: Metrological confirmation system for measuring equipment

ISO 10013:1995 *Ed. 1* *11 p. (F)* *TC 176 / SC 3*
Guidelines for developing quality manuals

03.120.20 Product and company certification. Conformity assessment

ISO/IEC Guide 7:1994 *Ed. 2* *3 p. (B)* *CASCO*
Guidelines for drafting of standards suitable for use for conformity assessment

ISO/IEC Guide 22:1982 * *Ed. 1* *4 p. (B)* *CASCO*
Information on manufacturer's declaration of conformity with standards or other technical specifications

ISO/IEC Guide 23:1982 * *Ed. 1* *4 p. (B)* *CASCO*
Methods of indicating conformity with standards for third-party certification systems

***(under revision)**

ISO/IEC Guide 25:1990 * *Ed. 3* *7 p. (D)* *CASCO*
General requirements for the competence of calibration and testing laboratories

ISO Guide 27:1983 * *Ed. 1* *5 p. (C)* *CASCO*
Guidelines for corrective action to be taken by a certification body in the event of misuse of its mark of conformity

ISO/IEC Guide 28:1982 * *Ed. 1* *16 p. (H)* *CASCO*
General rules for a model third-party certification system for products

ISO/IEC Guide 39:1988 * *Ed. 2* *8 p. (D)* *CASCO*
General requirements for the acceptance of inspection bodies

ISO/IEC Guide 40:1983 * *Ed. 1* *3 p. (B)* *CASCO*
General requirements for the acceptance of certification bodies

ISO/IEC Guide 42:1984 *Ed. 1* *6 p. (C)* *CASCO*
Guidelines for a step-by-step approach to an international certification system

ISO/IEC Guide 43:1984 * *Ed. 1* *6 p. (C)* *CASCO*
Development and operation of laboratory proficiency testing

ISO/IEC Guide 44:1985 *Ed. 1* *13 p. (G)* *CASCO*
General rules for ISO or IEC international third-party certification schemes for products

ISO/IEC Guide 48:1986 * *Ed. 1* *9 p. (E)* *CASCO*
Guidelines for third-party assessment and registration of a supplier's Quality System

ISO/IEC Guide 53:1988 * *Ed. 1* *13 p. (G)* *CASCO*
An approach to the utilization of a supplier's quality system in third party product certification

ISO/IEC Guide 56:1989 *Ed. 1* *4 p. (B)* *CASCO*
An approach to the review by a certification body of its own internal quality system

ISO/IEC Guide 57:1991 * *Ed. 2* *3 p. (B)* *CASCO*
Guidelines for the presentation of inspection results

ISO/IEC Guide 58:1993 *Ed. 1* *6 p. (C)* *CASCO*
Calibration and testing laboratory accreditation systems — General requirements for operation and recognition

ISO/IEC Guide 60:1994 *Ed. 1* *2 p. (A)* *CASCO*
ISO/IEC Code of good practice for conformity assessment

ISO/CASCO 226 (Rev. 2) Draft – General requirements for assessment and accreditation of certification/registration bodies **(will be published as ISO/IEC Guide 61)**

ISO/CASCO 227 (Rev. 2) Draft – General requirements for bodies operating assessment and certification/registration of quality systems **(will be published as ISO/IEC Guide 62)**

***(under revision)**

03.120.30 Application of statistical methods

ISO 2602:1980 *Ed. 2* *5 p. (C)* *TC 69*
Statistical interpretation of test results — Estimation of the mean — Confidence interval

ISO 2854:1976 *Ed. 1* *46 p. (T)* *TC 69*
Statistical interpretation of data — Techniques of estimation and tests relating to means and variances

ISO/DIS 2859-0 *Ed. 1* *56 p.* *TC 69 / SC 5*
Sampling procedures for inspection by attributes —
Part 0: Introduction to the ISO 2859 attribute sampling system (Revision of ISO 2859:1974 and of Addendum 1:1977)

ISO 2859-1:1989 *Ed. 1* *67 p. (V)* *TC 69 / SC 5*
Sampling procedures for inspection by attributes —
Part 1: Sampling plans indexed by acceptable quality level (AQL) for lot-by-lot inspection

Technical Corrigendum 1:1993 to ISO 2859-1:1989
Ed. 1 *2 p.* *TC 69 / SC 5*

ISO/DIS 2859-1 *Ed. 2* *87 p.* *TC 69 / SC 5*
Sampling procedures for inspection by attributes —
Part 1: Sampling plans indexed by acceptable quantity level (AQL) for lot-by-lot inspection (Revision of ISO 2859-1:1989)

ISO 2859-2:1985 *Ed. 1* *21 p. (L)* *TC 69 / SC 5*
Sampling procedures for inspection by attributes —
Part 2: Sampling plans indexed by limiting quality (LQ) for isolated lot inspection

ISO 2859-3:1991 *Ed. 1* *16 p. (H)* *TC 69 / SC 5*
Sampling procedures for inspection by attributes —
Part 3: Skip-lot sampling procedures

ISO 3207:1975 *Ed. 1* *15 p. (H)* *TC 69*
Statistical interpretation of data — Determination of a statistical tolerance interval

Addendum 1:1978 to ISO 3207:1975
Ed. 1 *3 p. (B)* *TC 69*

ISO 3301:1975 *Ed. 1* *6 p. (C)* *TC 69*
Statistical interpretation of data — Comparison of two means in the case of paired observations

ISO 3494:1976 *Ed. 1* *44 p. (S)* *TC 69*
Statistical interpretation of data — Power of tests relating to means and variances

ISO 3534-1:1993 *Ed. 1* *53 p. (T)* *TC 69 / SC 1*
Statistics — Vocabulary and symbols —
Part 1: Probability and general statistical terms

ISO 3534-2:1993 *Ed. 1* *39 p. (Q)* *TC 69 / SC 1*
Statistics — Vocabulary and symbols —
Part 2: Statistical quality control
Bilingual edition

ISO 3534-3:1985 *Ed. 1* *33 p. (Q)* *TC 69 / SC 1*
Statistics — Vocabulary and symbols —
Part 3: Design of experiments
Bilingual edition

ISO 3951:1989 *Ed. 2* *107 p. (XA)* *TC 69 / SC 5*
Sampling procedures and charts for inspection by variables for percent nonconforming

ISO/DIS 5479 *Ed. 1* *22 p.* *TC 69 / SC 6*
Statistical interpretation of data — Tests for departure from the normal distribution

ISO 5725-1:1994 *Ed. 1* *17 p. (J)* *TC 69 / SC 6*
Accuracy (trueness and precision) of measurement methods and results —
Part 1: General principles and definitions

ISO 5725-2:1994 *Ed. 1* *42 p. (S)* *TC 69 / SC 6*
Accuracy (trueness and precision) of measurement methods and results —
Part 2: Basic method for the determination of repeatability and reproducibility of a standard measurement method

ISO 5725-3:1994 *Ed. 1* *25 p. (M)* *TC 69 / SC 6*
Accuracy (trueness and precision) of measurement methods and results —
Part 3: Intermediate measures of the precision of a standard measurement method

ISO 5725-4:1994 *Ed. 1* *23 p. (L)* *TC 69 / SC 6*
Accuracy (trueness and precision) of measurement methods and results —
Part 4: Basic methods for the determination of the trueness of a standard measurement method

ISO 5725-6:1994 *Ed. 1* *41 p. (S)* *TC 69 / SC 6*
Accuracy (trueness and precision) of measurement methods and results —
Part 6: Use in practice of accuracy values

ISO 7870:1993 *Ed. 1* *9 p. (E)* *TC 69 / SC 4*
Control charts — General guide and introduction

ISO/DTR 7871 *Ed. 1* *TC 69 / SC 4*
Guide to quality control and data analysis using cusum techniques

ISO 7873:1993 *Ed. 1* *13 p. (G)* *TC 69 / SC 4*
Control charts for arithmetic average with warning limits

ISO 7966:1993 *Ed. 1* *21 p. (L)* *TC 69 / SC 4*
Acceptance control charts

ISO 8258:1991 *Ed. 1* *29 p. (N)* *TC 69 / SC 4*
Shewhart control charts

Technical Corrigendum 1:1993 to ISO 8258:1991
Ed. 1 *1 p.* *TC 69 / SC 4*

ISO 8422:1991 *Ed. 1* *45 p. (S)* *TC 69 / SC 5*
Sequential sampling plans for inspection by attributes

Technical Corrigendum 1:1993 to ISO 8422:1991
Ed. 1 *1 p.* *TC 69 / SC 5*

ISO 8423:1991 *Ed. 1* *39 p. (R)* *TC 69 / SC 5*
Sequential sampling plans for inspection by variables for percent nonconforming (known standard deviation)

Technical Corrigendum 1:1993 to ISO 8423:1991
Ed. 1 *1 p.* *TC 69 / SC 5*

ISO/TR 8550:1994 *Ed. 1* *48 p. (Q)* *TC 69 / SC 5*
Guide for the selection of an acceptance sampling system, scheme or plan for inspection of discrete items in lots

ISO 8595:1989 *Ed. 1* *3 p. (B)* *TC 69 / SC 1*
Interpretation of statistical data — Estimation of a median

ISO/DIS 11095 *Ed. 1* *TC 69 / SC 6*
Linear calibration using reference materials

ISO/DIS 11453 *Ed. 1* *TC 69 / SC 3*
Statistical interpretation of data — Tests and confidence intervals relating to proportions

ISO/DIS 11843-1 *Ed. 1* *TC 69 / SC 6*
Capability of detection —
Part 1: Terms and definitions

ISO CENTRAL SECRETARIAT

1, rue de Varembé
Case postale 56
CH-1211 Genève 20
Switzerland
Telephone: + 41 22 749 01 11

Sales Department

Telefax: + 41 22 734 10 79
Email:
Internet: SALES@ISOCS.ISO.CH
X.400: C=CH; ADMD=ARCOM; PRMD=ISO; O=ISOCS; S=SALES

Addresses of Authors

M.J. Allison
Stockdales Pty. Ltd
Consultants to Industry
Brighton
Victoria 3086
AUSTRALIA

Laura Alvarez
National Center for Metrology
CENAM
Aptdo. Postal 1-100
Querétaro
Mexico 76000

David L. Berner
AOCS
1608 Broadmoor Drive
Champaign
Illinois
61826-3489 USA

Trean Korbelak Blumenthal
Libra Technologies, Inc
16 Pearl Street
Metuchen,
NJ 0884
USA

Peter Bowron
Toxicology Unit
Royal North Shore Hospital
Macquarie Hospital Campus
Badajoz Rd., North Ryde
N.S.W. AUSTRALIA

Leslie Burnett and Judy Banning
Department of Clinical Chemistry
Institute of Clinical Oathology and
Medical Research
Westmead Hospital
Westmead NSW 2145 AUSTRALIA

Gavan J. Canavan and Michael J. Liddy
Victoria Forensic Science Centre
Victoria Police
Forensic Drive
Macleod Victoria
AUSTRALIA 3085

J.M. Christensen
Department of Chemistry and Biochemistry
National Institute of Occupational Health
Lersoe Parkallé 105
DK-2100 Copenhagen
DENMARK

Carl A. Gibson and Len R. Stephens
Victorian Institute of Animal Science
475 Mickleham Road
Attwood
Victoria 3049
AUSTRALIA

E.M. Gibson
Environmental Chemistriy Unit
Environment Protection Authority
GPO Box 4395 QQ
Melbourne Victoria
AUSTRALIA 3001

David N. Gidley and Michael J. Liddy
Victoria Forensic Science Centre
Victoria Police
Forensic Drive
Macleod Victoria
AUSTRALIA 3085

Pierre M. Gy
Sampling Consultant
14, Avenue Jean-de-Noailles
06400 Cannes
FRANCE

J.P. Hammond and M.J. Long
UV Marketing
ATI Unicam
York Street
Cambridge
CB1 2PX, UK

A.M. Henderson
B. Met. Eng.
PO Box 32
POINT LONSDALE
VIC 3225
AUSTRALIA

J.W. Hosking
Chemistry Centre (WA)
125 Hay St.
East Perth
Western
AUSTRALIA 6004

I.R. Juniper
C Chem, MRSC
NATA Australia
7 Leeds Street
Rhodes
NSW 2138 AUSTRALIA

Bernard King
Laboratory of the Government Chemist
Queens Road
Teddington - Middlesex
TW11 OLY
UK

Harry Klich
Bundesanstalt für Materialforschung
Rudower Chaussee 5

12489 Berlin
GERMANY

Jesper Kristiansen
Department of Chemistry and Biochemistry
National Institute of Occupational Health
Lersoe Parkallé 105
DK-2100 Copenhagen
DENMARK

Daniel Kwok
Crosby Associates
Level 8
2 Elizabeth Plaza
North Sydney
NSW 2066 AUSTRALIA

Zhao Min and Han Yongzhi
National Research Center for CRMs
Beijing
P.R.
CHINA

Alfredo M. Montes Nino
Xenobioticos SRL
Bolivia 5826/28
Buenos Aires (1419)
ARGENTINA

Anne Nicholson
Anne Nicholson Enterprise Pty Ltd
32 Millicent Street
Rosana VIC
AUSTRALIA 3084

Erik Olsen and Frands Nielsen
Department of Occupational Hygiene
National Institute of Occupational Health
Lerso Parkallé 105
DK-2100 Copenhagen
DENMARK

Laszlo Paksy
Metalcontrol Kft. Miskolc
Vasgyari u 43
H-3540
HUNGARY

M. Parkany
ISO Central Secretariat
Geneva 20
Case postale 56
CH-1211 SWITZERLAND

A.J. Poynton
Kraft Foods Ltd
PO Box 1673N
Melbourne 3000
AUSTRALIA

S.D. Rasberry
National Institute of Standards and Technology
Gaithersburg
MD 20899
USA

W.G. de Ruig and H. van der Voet
DLA
State Institute for Quality Control of
Agricultural Products (RIKILT-DLO)
PO Box 230
6700 AE Wageningen
NETHERLANDS

Maritta Siloaho and Eino Puhakainen
Department of Clinical Chemistry
Kuopio University Hospital
FI-70211 Kuopio
FINLAND

Alan L. Squirrell
Mnager, Proficiency Testing
National Association of Testing Authorities
7 Leeds Street
Rhodes NSW 2138
AUSTRALIA

Adam Uldall
Department of Clinical Chemistry
Herlev University Hospital
DK-2730 Herlev
DENMARK

S.M. Weeks and N.G.C. Bruer
The Australian Wine Research Institute
PO Box 197
Glen Osmond
South
AUSTRALIA 5064

P.H. Wright
Gas Grid Quality Assurance Branch
Engineering Services Department, Gas & Fuel
1136 Nepean Higway
Highett, Victoria
AUSTRALIA

C.M. van Wyck
ACIRL
Junction Street
Telarah
NSW 2320
AUSTRALIA

Subject Index

A

B

C

D

E

F

G

H

I

J

K

L

M

N

O

P

Q

R

S

T

U

V

W

Y

Z